Science for Responsible Citizenship

Science for Responsible Citizenship helps the reader to understand how science works and how science can help us develop reasonable and logical solutions to worldwide problems. Basic science knowledge and possible solutions based on scientific reasoning are explored, with a focus on interdisciplinary topics including disease transmission, energy, and climate change. The book starts with discussing the role of science in modern society, before tackling the basic science of global warming, human health, chemical processes that contribute to life on Earth, and the importance of water. The second half of the book describes the science of the electrical grid, nuclear reactions, devices and reactors, the Earth's habitability, electronics, and optics, and concludes with a look ahead at the scientific and technological challenges of future energy needs and sustainable energy. This book can be used as part of a general science course at the college level and is also accessible to anyone with a high school background.

Key Features:

- Addresses the fundamentals of scientific knowledge and reasoning in a clear and accessible style.
- Offers a timely exploration of several existential challenges that can be addressed by science.
- Provokes and encourages thought on how to solve some of our global problems.

Blane Baker is Professor of Physics at his alma mater, William Jewell College, where he returned to teach in 1999. He regularly teaches courses in general physics, electronics, and quantum mechanics, along with a popular sport science course for nonscience majors. Baker is an active contributor to the American Association of Physics Teachers (AAPT) and served as co-chair of the Sigma Pi Sigma 2022 Physics Congress. His areas of interest include electronics, sustainable energy, and materials science. He is a member of the American Physical Society (APS), American Association of Physics Teachers (AAPT), Society of Physics Students (SPS), and Sigma Pi Sigma. He also holds the Wallace A. Hilton Endowed Chair in Physics at William Jewell College.

Science for Responsible Citizenship

Blane Baker

CRC Press
Taylor & Francis Group
Boca Raton London New York

CRC Press is an imprint of the
Taylor & Francis Group, an **informa** business

Designed cover image: Shutterstock

First edition published 2025
by CRC Press
2385 NW Executive Center Drive, Suite 320, Boca Raton FL 33431

and by CRC Press
4 Park Square, Milton Park, Abingdon, Oxon, OX14 4RN

CRC Press is an imprint of Taylor & Francis Group, LLC

© 2025 Blane Baker

ISBN: 9781032841595 (hbk)
ISBN: 9781032832180 (pbk)
ISBN: 9781003511472 (ebk)

DOI: 10.1201/9781003511472

Typeset in Minion Pro
by Newgen Publishing UK

Contents

CHAPTER 1 ▪ Science and Its Role in Modern Society   1

CHAPTER 2 ▪ Global Warming: Crisis and Opportunity   11

CHAPTER 3 ▪ Human Disease and Health   26

CHAPTER 4 ▪ The Chemical Medley   37

CHAPTER 5 ▪ The Electrical Grid: Boom or Bust   48

CHAPTER 6 ▪ The Nuclear Dilemma   56

CHAPTER 7 ▪ Our Place in the Universe   71

CHAPTER 8 ▪ Electronics in the 21st Century   81

CHAPTER 9 ▪ Our Energy Future   90

INDEX, 106

CHAPTER 1

Science and Its Role in Modern Society

Discoveries within the natural sciences continue to inform our understanding of the world around us. In addition to increasing our knowledge, efforts in the sciences contribute to the development of numerous technologies, including ones in communications, transportation, and computing and graphics. To better understand the nature of science and its impacts on humanity, the following sections provide an overview of science, its methods and limitations, and examples of two scientific theories that describe how matter behaves. Discussion questions focus on stimulating further thought on the role of science in society, the need for scientists to promote authentic engagement with the public, and the most effective ways to solve interdisciplinary problems.

SCIENCE AS A DISCIPLINE

From the root word *scientia*, meaning "knowledge," the natural sciences are disciplines that seek to understand how nature works. Methods of science focus on obtaining knowledge by asking scientific questions, gathering and interpreting data, developing hypotheses, and testing them with ongoing experiments. While no one specific method is appropriate for every researcher, scientific studies generally have common strategies.

DOI: 10.1201/9781003511472-1

Scientists often begin by examining the results of previously unexplained experiments, novel observations, or unanswered questions. Researchers then develop hypotheses to explain these results and design and conduct experiments to test the hypotheses. Laboratory measurements, computer simulations, or theoretical calculations yield valuable data. After analysis and interpretation, scientists determine if the working hypotheses are valid.

With each new confirmation, hypotheses gain credibility and ultimately contribute to theories or physical laws. Hypotheses that do not meet the standard of prediction are either modified to explain the new results or discarded entirely. Even well-established theories are subject to correction (or rejection) if new results show that they are inadequate.

Once established, scientific theories (and laws) serve as our best descriptions of nature. Critics often argue that theories are invalid, because they lack vetting and confirmation. Scientifically speaking, however, these arguments are clearly incorrect. In fact, scientific theories are the most valid and reliable explanations available. Rigorous experimentation and subsequent refinements of theories increase their validity over time. In many instances, reliable theories require hundreds of years of investigations to develop. Two theories that explain many aspects of how matter behaves are discussed in a section below.

Studies in the natural sciences encompass every innate part of the universe, ranging from the tiniest known particles (quarks) to the universe itself. Scientists generally focus on elucidating the laws that govern nature. As part of their repertoire for gaining knowledge, investigators ask scientific questions. Such questions specify what we seek to understand by the methods of science. A question such as, "How does a rainbow form?" is within the realm of science. Whereas a question like, "How should I behave in this situation?" is not. Moral questions generally fall under the aegis of philosophy and ethics.

As a human endeavor, science has intrinsic limitations based on both boundaries and methodologies. Individual disciplines generally study certain systems within nature. As an example, physics seeks to understand the behavior of material systems (including energy) within the natural world. Questions of human morals, actions, and behaviors generally fall within other disciplines such as philosophy, psychology, and sociology.

Despite established boundaries, the natural sciences often inform other areas of study. For example, when social scientists study how someone responds during a pleasurable experience, natural scientists enhance that knowledge by monitoring chemical, electrical, and magnetic effects within the body and brain. As techniques and methods progress, boundaries within science naturally evolve over time. For instance, with the development of computational biology, investigators now have capabilities to model biological data and contribute to solving biological and biomedical problems.

In addition to boundaries related to scope, scientific endeavors have limitations based on what methodologies are available at the time. Methodologies evolve so that those beyond our grasp 20 or 30 years ago are available today, thus allowing scientists access to more techniques for studying nature. A recent example is the detection of cosmic gravitational waves in 2015 using (LASER) light interference methods. Related versions of this technique dating back to 1887 showed that light travels at the same speed in all directions in all inertial frames. In these kinds of experiments, a light beam is split into two beams that travel along perpendicular paths, reflect off mirrors, and then recombine to produce known interference patterns. If the length of either path or the speed along either direction changes, phase differences between the two beams occur, and the interference pattern changes in a predictable way.

To make these techniques sensitive enough to detect the effects of gravitational waves on space itself, the two LASER beams move along 4.0 km paths within tubes evacuated to one-billionth of an atmosphere, thus greatly reducing interactions between light and air molecules. If a gravitational wave interacts with the detector, it creates a ripple in space along one of the directions of the two beams. This ripple alters the distance that light travels in that direction, thus changing the original interference pattern. Observations of such changes confirm that a gravitational wave passed the detector.

When scientists first considered gravitational wave detection, the apparatus described above was thought impossible. Many technological challenges required novel solutions to bring these experiments to fruition, including development of vacuum technology for very large cavities, honing of interference techniques to detect a change in distance of 0.0001 the width of a proton, and even construction of the LASER beam lines to account for the curvature of the Earth.

For the case here humankind developed new technologies to overcome the obstacles that previously prevented the detection of gravitational waves.

Scientists conceived of how they might accomplish these tasks many years ago, but the methods for detection were not feasible until about 2015–2016. Other methodologies are not yet in the conception stage. For example, knowledge of the cosmic state before the Big Bang is not yet accessible in theory or practice.

EXAMPLES OF TWO SCIENTIFIC THEORIES OF MATTER

Modern scientific thought, founded on observation, experiment, and measurement, began in the 1500s. Today, hundreds of scientific theories abound, providing us with the most compelling explanations of how various systems in nature work. The discussion below describes two theories—the atomic theory of matter and the theory of universal gravitation—both of which are cornerstones of how we understand the behavior of matter. These theories emerged over many years of devoted thought and experimentation by a host of scientists.

The theory that all matter consists of particles called atoms is foundational to our understanding of nature. Many scientists, including the likes of Richard Feynman, believe that the atomic theory is the most important knowledge acquired by humans. Elaboration of this theory explains an enormous number of different phenomena, ranging from the behavior of gases to bonding of atoms.

Essential to this theory is the notion that atoms are in perpetual motion. (Theoretically, atoms maintain motion even at absolute zero according to quantum-mechanical models.) Motions of atoms are crucial for describing many of the properties of matter. Consider, for instance, a gas confined to a cylinder with a piston inserted into its open end. Once inserted, the piston settles at some equilibrium point along the length of the cylinder. Equilibrium occurs when the weight of the piston and the forces imparted by the gas molecules within the cylinder balance one another.

Molecules interacting with surfaces produce forces due to collisions. In an ideal, one-dimensional collision, a molecule strikes a surface head-on and returns with the same speed as it struck the surface. This kind of collision, characterized as elastic, obeys the law of conservation of momentum like all collisions in nature. During elastic collisions, the speed of the particle is the same before and after collision, but the particle moves in the opposite direction. As a result, the molecule experiences an acceleration and thus a force by Newton's second law ($F = ma$). The force on the molecule is due to interactions between the molecule and the solid surface. Not only does the surface exert a force on the molecule, but the molecule also exerts a force on the surface by Newton's third law. An equivalent way to characterize the force is to determine the change in momentum of the particle during the duration of the collision. The ratio of the change in momentum to the time interval is also a measure of the force.

A force exerted over an area determines the pressure of a system. The pressure of an ideal gas depends on the number of molecules present, the volume of the system, and the temperature of the gas. The molecular theory of gases explains these factors. As the number of gas molecules (particles) increases, the pressure increases (linearly) due to a larger number of molecules striking the container within each time interval. During collision, each molecule experiences a change in momentum; thus, the overall change in momentum in each unit of time increases with a larger number of molecules. A greater change in momentum in a unit of time produces a larger force.

When the number of gas particles and temperature are constant, an increase in the volume of the gas reduces the pressure. Here, similar arguments hold from above. As the volume increases, the number of particles present in each unit volume decreases. Thus, fewer particles strike a given surface area (per unit of time) and the average force and pressure decrease.

The dependence of pressure on temperature originates from energy relationships. As the temperature of a gas increases, the average kinetic energy of the constituent particles within the system increases. Kinetic energy depends on both the mass of a moving particle and its speed. When a particle strikes a surface with greater speed, the force imparted by the particle is larger (provided the collision time remains about the same). A larger force also results in greater pressure.

Atomic theory is also critical to explain the formation of bonds between atoms. Bonding results from electron (particle) interactions between one or more atoms. In some cases, atoms share electrons, and, in other cases, they transfer electrons from one to another. The number of electrons that participate in bonding partially determines how atoms combine in certain ratios to form molecules. When one oxygen atom combines with two hydrogen atoms, a water molecule forms.

Another fundamental property of matter is that every particle (with mass) attracts every other particle in the universe with a gravitational force. This force depends linearly on the mass of each object and varies inversely with the square of the separation between the two masses. The equation takes the form:

$$F = G\frac{m_1 m_2}{r^2}$$

where G is a universal constant, m_1 and m_2 are the two masses, and r is the separation between the masses.

Before discoveries of how gravity behaves between massive objects, early Greeks observed the effects of gravity but attributed them to systems seeking their natural places in the universe. A stone naturally falls to the ground as it returns to its origin (the Earth). Smoke rises because it seeks its place in the sea of air surrounding the Earth. While these ideas persisted for hundreds of years, they had no basis in scientific thought and/or experiment.

Early clues that massive objects experience forces stemmed from the existence of regular motions of planets within the solar system. Not only do planets move relative to background stars, but they also appear in the same regions of the sky with predicted regularity. Such motions suggest forces keeping the objects in regular orbits, akin to those of a mass attached to a string moving in a circle as someone swings it. To discern why these orbits occur, the astronomer Tycho Brahe made numerous observations of planets over many years and recorded these results in voluminous tables.

Using Tycho's observational data, Johannes Kepler developed three laws of planetary motion. The first one says that planets follow elliptical paths with the Sun located at one of the foci. An ellipse is a closed curve shaped like an oval and formed by elongation of a circle along one axis. More formally, an ellipse consists of a set of points such that the sum of the distances from any point on the curve to two fixed points, called foci, is always the same.

Kepler also observed that planets move at varying speeds along their paths around the Sun. Observations confirm that orbital speeds are faster when planets are closer to the Sun and slower when planets are farther away. Quantitatively, variations in speed with distance translate to planets sweeping out equal areas in equal times. To determine areas of orbits, imagine a line segment from the Sun to the path of the planet at a certain time and then imagine a second segment after a fixed time later. The two segments and the arc of the path define an area for a fixed time interval.

Comparisons along any part of the path (for a fixed time interval) reveal that the areas swept out by the orbiting body are always equal.

The third law describes how the period of the planet's path depends on the size of the orbit. The mathematical relation says that the period is proportional to the 3/2 power of the orbit's size. The period here is the time interval required for the planet to go completely around the Sun; the size of the orbit is the length of the semimajor axis of the planet's elliptical orbit. The 3/2 power relation says that if the distance increases by a factor of 4, the period increases by a factor of 8. For reference the Earth is at 1 astronomical unit (AU) from the Sun and takes 1 year to complete a full path around the Sun. By contrast, Jupiter, located at 5.2 AU from the Sun, takes 12 years to orbit.

Kepler's work on planetary orbits coincided with Galileo's experiments on motions of objects such as spheres rolling along inclined planes. From numerous observations, Galileo concluded that objects at rest or in motion tend to maintain those states of motion due to a universal property of matter known as inertia. Newton modified this explanation to say that a change in the state of momentum (motion) of an object requires a net force. A more modern version of this property states that an object at rest tends to stay at rest, and an object in motion tends to stay in motion unless acted upon by a net external force.

To explain the motions of planets, Newton and others suggested that a sideways force acts on the bodies to cause them to move in closed and regular orbits. As outlined in the law of inertia, the motion of any object (including a planet) continues along a straight-line path when left undisturbed. A sideways force acting toward a central point continuously changes the planet's direction of motion, thus causing it to move in a circular (or elliptical) path. Indeed, observed motions of planets follow these predictions.

Further investigations suggested that the sideways force needed to maintain closed planetary orbits acts toward the Sun. These results and observations of small bodies (moons) orbiting Jupiter provided Newton with sufficient evidence to theorize that attractive gravitational forces occur among all massive bodies in the universe.

Using an elegantly designed balance, Cavendish conducted experiments to confirm the universal nature of gravity and measure the universal constant G. The specialized balance consists of two objects, each with mass m, attached to the ends of a rod, suspended by a fine fiber or thin metal wire. Experiments place two larger masses (each with mass M) near the smaller masses to cause deflection of the suspended rod. Measurements of the deflection confirm the nature of universal gravitation and provide a way to determine G.

For ordinary massive objects the classical laws of gravity work exceedingly well and allow scientists to make many accurate predictions. However, when objects move at high rates of speed, relativistic effects occur. In modifications to the classical theory of gravity, energy and matter are interchangeable according to Einstein's work in relativity. As a result, light waves possessing energy experience gravitational forces as they pass near massive objects. These forces cause light from distant objects (such as stars) to bend as their paths come close to objects such as the Sun. During solar eclipses scientists observe displacement of stars relative to their true locations due to gravitational effects on light. These measured displacements agree with relativistic treatments of gravity.

As illustrated above, all theories in science are subject to scrutiny. Working theories that conflict with new scientific data warrant modification (or even elimination). Thus, science operates as a self-correcting endeavor, seeking to obtain accurate knowledge of the natural world.

DISCUSSION QUESTIONS

What Is the Role of Science in Society?

The most important role that science plays in society is to generate new knowledge. Scientists pursue knowledge by asking relevant questions and seeking to answer them through established methods. Scientific investigations generally focus on pursuing research projects that gather and analyze data, provide scientific interpretations of that data, and contribute to theoretical explanations of nature.

In conjunction with research, scientists communicate their work to peers for evaluation and review. Dissemination of results occurs through meetings (and conferences) where workers present talks or posters and in peer-reviewed research publications. In these settings, other scientists offer criticism and encouragement. Presenting research results to others in the field and responding to critiques are necessary and valuable for the advancement of science.

Scientific knowledge is also important as a public resource. Scientific discoveries provide the needed technological background to develop new products, inform our policy decisions, and guide us in making wise personal choices. Policy decisions are complex, but scientists have the expertise to contribute needed scientific background and to model critical reasoning to inform decisions. Equipped with information and ways to think about the implications of actions, an informed citizenry can then advocate for policies that solve our most pressing problems.

Why are Citizens Skeptical of Scientists?

Trust in scientists continues to decline in the US. In 2023, surveys reported by the Pew Research Center indicated only 57% of Americans say that science has had a mostly positive effect on society, compared to 2019 when 73% answered affirmatively to this question. Citizens often doubt science, because scientific claims change over time—contrary to what people expect of science. While the evolving nature of science presents challenges, dismissal of scientific results fails to understand how science works and how science corrects itself, based on the latest evidence. The self-correcting process of science provides us with the best possible data and conclusions. Another reason cited for lack of trust in science is that it conflicts with people's social identities. Dissonance between one's belief system and evidence from science is often uncomfortable. When these kinds of conflicts occur, honest conversations provide avenues for living with these tensions.

In response to the latter argument, science does not promote certain political or social agendas. In fact, scientists have ethical obligations to present unbiased scientific results. Here the most helpful and honest approach is to clearly and accurately present the full body of scientific results and explain the most viable interpretations. As individuals, however, scientists have personal opinions and have the right to express those opinions. Solutions to problems ultimately require input from a wide range of constituencies. As citizens, our best approach is to advance policies that address multiple perspectives and promote the common good.

In working toward benefical solutions , conversations between the public and scientific experts are crucial. The public benefits by learning about exciting and innovative scientific research while scientists benefit by getting to know the public and learning their interests. Over time these relationships build trust. As trust develops, the public tends to consider scientific results more seriously. When the public lacks trust, however, they tend to exhibit cynicism and ill feelings. For many of us, trust develops by knowing a person and observing their actions over time.

What Are Effective Strategies for Solving Monumental Problems?

Solving difficult problems requires collaboration. As a current example, one of our most daunting problems is how to meet the energy demands of humankind in the context of diminishing resources and climate change. Here solutions must foster seamless incorporation of more sustainable energy, promote economic growth, and minimize harm to humans. Strategies and policies require engagement in primary research, application of results from fundamental studies to develop new technologies, development of infrastructure needed to distribute energy, and implementation of economic models to ensure stability. For example, fundamental research on materials for the solar cell and electrical transmission industries helps designers and engineers produce more efficient solar cell devices and develop the next generation of electrical transmission lines. Engineering, utility, and construction experts design and build the required infrastructure to transport energy efficiently. Business professionals contribute by providing new products at competitive prices, marketing those products, and developing economic incentives to convert to new technologies.

Collaboration is essential for addressing the global challenges that confront us. In that spirit, scientists must use their skills in critical thought to inspire confidence in scientific results, contribute to viable solutions, and promote a sustainable world.

CHAPTER 2

Global Warming

Crisis and Opportunity

Discussions of climate change—including environmental impacts, emergence of new weather patterns, and increased prevalence of natural disasters—often cause angst and confusion. While our instinctive reactions tend toward fear, denial, and even cynicism, a more proactive approach focuses on learning as much as possible about the current situation and taking actions to improve our environment.

The abilities of humans to reverse emerging trends are real and achievable. By the end of the 1980s, the use of chlorofluorocarbons (CFCs) had decreased ozone levels in the atmosphere and threatened to do more damage. With no clear solution at hand, various constituencies including businesses, governmental agencies, and citizens rallied to address this threat. Restrictions on CFCs, used in many products at that time, worked to reduce CFC levels in the atmosphere and ultimately led to restoration of ozone in the upper atmosphere.

Today's climate crisis is even more dire considering the magnitude of the threat and disagreements on how to act and on what time scale. However, through educational initiatives, activism, and honest conversations, folks now understand the need to respond. Each of us has daily opportunities to take personal actions, encourage others, and advocate for businesses and

DOI: 10.1201/9781003511472-2

others to adopt sustainable practices. Every proactive step taken toward mitigating climate change has an impact.

The current state of our climate is complicated, but the overall trends are clear. According to NOAA, the last ten years (2014–2023) are the warmest on record; moreover, several high-temperature records occurred in 2024. Levels of CO_2 continue to rise despite significant investments in renewable energy by a host of countries worldwide. To reverse the trends of hotter and statistically more severe weather, CO_2 levels must decrease. As our experiences have shown, our health, livelihoods, and even existence are at stake.

The following sections examine the science associated with climate change, consequences of climate change, skeptical views of climate change, possible solutions to the dilemma, and questions for discussion.

INTERACTIONS OF EM WAVES WITH MATTER AND CLIMATE CHANGE

Interactions of light (EM waves) with matter are crucial to understanding the nature of the greenhouse effect (and ultimately climate change). Electromagnetic (EM) waves consist of oscillating electric and magnetic fields traveling outward from a source. EM waves move at the speed of light and include many familiar types: visible light, radio waves, microwaves, X-rays, and infrared waves.

To visualize EM waves, imagine a rope tied to a support at one end and moved up and down on the other end. As the rope oscillates, a series of waves (each consisting of a full up-and-down cycle) propagate outward from the source (your hand). Descriptions of waves include the quantities amplitude, wavelength, frequency, and wave speed.

The amplitude of a wave is a measure of how far the oscillation moves above and below its starting point. Wavelength refers to the distance between one positive (or negative) peak and the next one; frequency is a measure of how many complete oscillations (or cycles of oscillation) occur each second. Wave speed refers to how much distance a wave covers in a fixed amount of time (usually 1 s). The product of the wave's frequency and its wavelength gives the wave speed. A wave's energy depends on its frequency; waves with higher frequencies have higher energies.

When EM waves interact with matter one or more processes generally occur. Reflection refers to interactions by which light returns to the medium from which it arrives. (For example, high class mirrors return approximately 99.999% of visible light upon reflection.) During reflection, the EM waves only penetrate the material by fractions of wavelengths, thus the incoming waves do not significantly excite atoms to produce absorption of the waves.

Absorption, in contrast with reflection, is the process of transferring energy to atoms and molecules via EM waves. In the visible light region, EM energies closely match those needed by electrons in lower energy states to jump to higher energy states. During absorption EM waves transfer energy to electrons, thus causing them to transition from lower to higher energy levels. Emission refers to electrons making transitions back to lower energy states. During this process the atoms or molecules release energy in the form of packets or bundles known as photons. Production of many photons that propagate through space results in radiation of EM waves.

In the infrared (IR) region, EM energies closely match those needed by molecules to initiate various kinds of vibration. When IR waves interact with molecules like CO_2, they absorb energy, causing them to transition from lower vibrational states to higher ones. These states correspond roughly to steps on a staircase where higher steps have higher energies. Once a molecule is in a higher state, it transitions back to a lower state by emitting EM waves of the same energy that it absorbed. This process is known as emission, because the molecule jumps to a lower energy state and emits EM radiation (generally IR here) as the transition occurs.

Common IR absorbers include water, CO_2, and methane. Because these molecules readily absorb in this region of the spectrum, they effectively trap IR within the lower atmosphere of the Earth, thereby creating a greenhouse effect. The greenhouse effect involves the interplay of several mechanisms: transmission, reflection, absorption, and radiation.

When EM waves (namely visible, ultraviolet, and infrared) from the Sun interact with the Earth's atmosphere, approximately 30% of that energy returns to space due to reflection. The remaining incident energy falls on the Earth's surface and warms it. The Earth then radiates energy as IR waves into the atmosphere so that some of the energy passes through the atmosphere and some undergoes absorption and re-radiation in all directions by greenhouse gases (including water vapor). Part of this energy returns to the Earth and is the origin of the greenhouse effect. A quantity known as radiative forcing is the net increase in downward radiation at the tropopause. As greenhouse gas concentrations rise, the amount of radiative forcing increases, causing increased warming effects.

TABLE 2.1   Major greenhouse gases under the Kyoto Protocol

Greenhouse gas
Carbon dioxide
Methane
Nitrous oxide
Hydrofluorocarbons
Perfluorocarbons
Sulfur hexafluoride
Nitrogen trifluoride

About 150 years ago John Tyndall performed several experiments, demonstrating the ability of gases such as CO_2 to absorb IR. Studies of these effects require an apparatus that passes light (often of varying wavelengths) through a tube of gas to determine the amount of light absorption. Absorption data for many gases are known from numerous experiments akin to the one described here. Gases of particular concern include those listed in Table 2.1.

VENUS: A RUNAWAY GREENHOUSE SCENARIO

The present-day climate of Venus demonstrates the results of a runaway greenhouse effect. The Earth and Venus have similar histories in that both planets evolved as rocky planets relatively near the Sun. During their early formation, both planets experienced significant volcanic activity, which released CO_2 into their atmospheres. On Earth, surface water (primarily oceans) readily absorbs CO_2, which combines easily with dissolved minerals to generate carbonate rocks. Today calcite and dolomite are the most common carbonates formed in the ocean. By contrast, Venus has no oceans to absorb CO_2, so all of it remains in the atmosphere.

The lack of oceans on Venus demonstrates how a runaway greenhouse effect depletes the oceans and produces extreme surface temperatures. Studies of Venus' evolution suggest the primitive planet probably supported oceans when the Sun's light was less bright than it is today. To illustrate the effects of greater intensity from the Sun, imagine the Earth located closer to the Sun. Less distance between the Sun and Earth causes surface temperatures to rise and more evaporation from oceans to occur. More water vapor then occupies the atmosphere, thus increasing the greenhouse effect. (Like CO_2, water is a greenhouse gas.) As atmospheric water vapor increases, surface temperatures rise due to enhanced radiative forcing. This cyclic effect (known generally as a positive feedback loop) continues to cause increased temperatures and depletion of surface water until either saturation of the atmosphere occurs or the oceans empty. (Once the oceans empty, carbonate rocks decompose, thus releasing additional CO_2 into the atmosphere.)

Currently, Venus is devoid of oceans (surface water) and has an atmosphere, composed primarily of CO_2. Mean surface temperatures are approximately 737 K, which are hot enough to melt lead. By contrast, Mars has some CO_2 on its surface and in its atmosphere; however, the surface atmospheric pressure on Mars is only 1% that of Earth and the amount of CO_2 is small in comparison to Venus. Thus, the greenhouse effect on Mars remains small. This small effect, together with the larger distance to the Sun keeps the mean surface temperature on Mars near 200 K.

OUTCOMES OF GLOBAL WARMING

The rise of CO_2 levels in the Earth's atmosphere currently shows no signs of decreasing, and, in fact continues to increase. Therefore, we expect to see even more pronounced effects due to this trend. The most obvious outcome is an enhanced greenhouse effect. Here the reasoning is simple: increased levels of CO_2 cause more radiative forcing of IR energy, thus more IR energy is present in the lower atmosphere.

Increases in energy in the atmosphere produce multiple effects. More energy present in the atmosphere warms the air and the oceans and drives additional weather events. The consequences of these changes include regional climate change, increased incidents of extreme weather, sea ice melting, and sea level rise. Several models predict global average surface temperature increases of a few degrees Celsius (°C) during this century, with more drastic increases in localized regions, particularly in the far north. Many secondary effects accompany the rise in temperatures.

One of the most obvious impacts is increased frequency of droughts and floods. Both occur due to more available energy and higher atmospheric and surface temperatures. Warmer temperatures drive more evaporation of water, tending to make dry places even drier. At the same time, more evaporation increases the availability of moisture available for rainfall. In the US the three highest-volume rainfall events over the past 70 years have occurred since 2016.

Drier and hotter conditions also lead to more wildfires due to several effects. Debris such as leaves, pine needles, and dead trees become extremely dry and thus easily combustible by lightning or careless hand-ling of campfires. In addition, these conditions usually produce hotter and more intense fires due to the lower water content of the fuel. Wildfires are difficult to extinguish and often burn until no more fuel is available.

As winters become warmer and shorter, many pests such as the pine beetle increase in population, thus leading to more devastation of forests due to these pests. Warmer overall temperatures reduce the incubation period of many diseases, transmitted by mosquitoes and other insects, and increase the geography of these pests. Thus, many diseases caused by mosquito-transmitted viruses such as dengue, chikungunya, West Nile, yellow fever, and Zika now occur in larger numbers and over broader geographic regions. Over time, increased temperatures cause melting of parts of the Earth's permafrost, containing trapped remains of plants. Such remains generally do not decay due to low year-round temperatures. If melting of the permafrost occurs, however, plant remains within the permafrost decay, thereby releasing additional CO_2 and methane.

The possibility that storms and severe weather events increase due to global warming requires careful consideration. For clarity it is impossible to tie an individual weather event—no matter how catastrophic—directly to global warming. A certain number of storms occur even if average global temperatures remain the same for many years. Weather events, like many other natural occurrences, are statistical in nature, so correlation does not guarantee causation.

Correlation between two events implies that the presence of one increases the likelihood of the second. For example, cirrhosis of the liver has statistical links to alcohol use, thus confirming correlation between the two. While the exact cause of cirrhosis in a particular case is not necessarily alcohol use, increased drinking leads to greater risk of developing cirrhosis.

By similar reasoning, increased concentrations of greenhouse gases result in more energy within the atmosphere. More energy in the atmos-phere contributes to increased likelihood of storms and severe weather events. Data collected by insurance companies reveal a general increase in the number of extreme weather and weather-related events from 1980 to the present. Other data indicate that within the 48 contiguous US states, the ratio of the number of record high temperatures to the number of record low temperatures increased from close to 1.0 in 1958 to over 2.0 in the 2000s.

Increasing surface temperatures naturally contribute to the melting of sea ice in regions like the Arctic Ocean. For reference, the average amount of sea ice has declined steadily since 1979. Melting of sea ice does not increase sea level but has other deleterious effects. Reduction of ice decreases the amount of light reflected by the Earth so that more absorption of energy occurs at the surface. Increased absorption of EM waves (energy) further increases surface temperatures, thereby amplifying global warming and melting of sea ice. This cycle of one effect triggering subsequent effects that reinforce the original change is a positive feedback loop. Such effects are alarming, especially if the original effect (global warming) is undesirable.

The decline of sea ice also leads to changes in the amount of salt in the water, referred to as salinity. In areas where ice melts, the local water salinity decreases due to the addition of fresh water. Such effects possibly trigger changes in ocean currents and reduce the productivity of fisheries. Both outcomes have the potential to impact local ecosystems, including food sources for humans.

Rising temperatures contribute to the increase of sea levels due to two main effects. When the temperature of a liquid (or solid) increases, constituent molecules within the system undergo increased motions and, therefore, tend to experience increased separations. When molecules attain larger (average) separations, the overall volume of the material increases. Expansion of bodies of water leads to a rise in sea level. Today sea level is approximately 21-24 cm higher than in 1880, primarily due to thermal expansion.

Another cause of sea level rise in the context of increasing temperatures is the melting of glaciers. Melting of glacial ice contributes to sea level rise as water from molten glaciers flows into the oceans. In addition, the decline of glaciers on land reduces the amount of reflection (or increases the amount of light absorption). This increase in absorption causes further increases in surface temperature and additional warming.

As atmospheric CO_2 levels rise, ocean waters increase in acidity due to absorption of CO_2 in the air. The carbon cycle characterizes the flow of carbon among systems on the Earth. In the absence of human influence, the flows of carbon among the air, plants, and oceans roughly balance overall. When human activities introduce additional carbon (CO_2) into the atmosphere, an imbalance between atmospheric and oceanic carbon occurs. Given this imbalance, oceans absorb large fractions of the atmospheric CO_2.

Absorption of CO_2 impacts acidity of the oceans due to reaction of water molecules with CO_2 to form carbonic acid. Once formed, carbonic acid readily dissociates to produce hydrogen ions, which increase the pH and thus acidity of the water. Scientific measurements indicate that ocean waters are more acidic than even 50 years ago.

SKEPTICAL VIEWS OF CLIMATE CHANGE

Skeptics continue to raise doubts about the existence of climate change and/or our abilities to mitigate such effects, despite extensive efforts to educate the public on the causes and consequences of climate change.

One of the common arguments of skeptics is that the Earth is not warming as expected (or that warming is slowing). One way to evaluate warming effects is to obtain the yearly average global temperature and compare it to previous years. Determining average global temperature is incredibly difficult given the need to average numerous local temperatures from around the globe. One of the limitations here is simply the number of data points (weather stations) available. Even today, sparse data exist for certain regions of the globe, and, in the past, even less data were available.

Despite these uncertainties in the actual average temperature, data clearly indicate that in comparison to the average temperature in the 20th century, yearly averages from at least 1980 to the present are above those values. Data reported in 2024 suggest warming in the US is increasing by 60% more than the worldwide average.

This increase in yearly temperatures is consistent with our knowledge of global warming. Carbon dioxide is a known greenhouse gas, human activities such as combustion of fossil fuels release CO_2, and levels of CO_2 within our atmosphere continue to rise along with global temperatures. The inevitable conclusion is that rising CO_2 levels (at least partially) lead to rising temperatures. As atmospheric concentrations of CO_2 increase, these effects become even more severe according to models and predictions, based on many scientific studies.

As some skeptics contend, a more detailed look at the data shows some slowing of increasing surface temperature, especially during the late 1990s and possibly beyond. This effect is somewhat mystifying given that rising levels of CO_2 trap more energy within the atmosphere, thus creating conditions for rising surface temperatures. While land-based temperature increases have slowed during certain periods, measurements indicate that warming of ocean water has accelerated during the same periods.

Additional studies show that glacial melting continues unabated, thus many trends associated with climate change persist.

Another skeptical view of climate change (global warming) says that the Earth is warming, but these changes are independent of human activities. One theory asserts that natural cycles associated with the Sun are the cause of warming. The Sun exhibits an 11-year cycle in which irradiance values vary in known ways. (Irradiance refers to how much energy per unit time per unit area strikes the Earth.) Additional modern data, however, reveal that surface temperatures on Earth continued to increase even during times in which solar irradiance decreased.

Beyond the arguments above, analyses of warming of the Earth's atmosphere due to solar activity predict temperature increases throughout the lower and upper regions of our atmosphere. Measurements, however, indicate warming of lower regions of the atmosphere and cooling of the upper layers. These effects are consistent with warming due to trapping of energy via greenhouse gases. Increased trapping of energy via greenhouse gases in the lower atmosphere reduces the amount of energy available to interact with the upper regions. While effects due to natural cycles of the Sun demand our attention, global increases in temperature show no relation with regular solar cycles. To date, scientific models of climate cannot predict observed temperatures without including human activities.

A third line of reasoning espoused by skeptics acknowledges human contributions to global warming but maintains that these effects are of no concern. Some even argue that warmer temperatures will positively impact humans by making some of the colder regions more tolerable. As part of their reasoning here, skeptics often point to the fact that CO_2 levels have varied for hundreds of thousands of years, and yet humans have survived. Indeed, historical data (from ice cores) clearly show correlations between rising (and falling) CO_2 levels and rising (and falling) temperatures. In the past, CO_2 levels exhibited peaks at around 300 ppm but later returned to levels around 200 ppm or less. (Each rise corresponded with a rise in temperature, and each decrease corresponded with a decrease in temperature.)

Several differences between the past and present are noteworthy. The current temperature of the Earth is close to the maximum attained over the past 800,000 years. In addition, current CO_2 levels are 40% higher than at any time during the 800,000-year interval and continue to rise. More dramatic increases in CO_2 levels and temperatures span only a few years, whereas historical changes occurred over centuries. The unprecedented recent changes trigger the effects discussed above.

The rise and fall of temperatures (and CO_2 levels) from the historical record are well understood and do not involve major human activities. For reference, the Industrial Revolution dates to about 1760, thus the range of human influence is less than 300 years. The periodic and natural warm periods and ice ages are well explained by the Milankovitch cycles. Periodic changes in the Earth's axis tilt and orbit due to gravitational effects generated by the Sun, Moon, and other planets produce known changes in the amount of EM radiation reaching us.

These cyclic changes alone cannot account for observed variations in historical temperatures (and climate); however, they trigger other effects that amplify temperature changes. A cycle often begins by slight warming of the Earth due to the orbital effects described. As the Earth warms, ocean waters release increased amounts of CO_2 that through the greenhouse effect cause additional warming. With these increases in temperature, more evaporation occurs, thus causing more water vapor to enter the atmosphere as a greenhouse gas. Through these positive feedback mechanisms, the Earth warms more than expected from the orbital effects alone. During natural cooling cycles, less greenhouse emissions occur, leading to amplified cooling. Both warming and cooling exhibit amplification due to triggering effects associated with natural variations in temperature. While these natural cycles produce temperature changes, they cannot fully explain temperature changes observed today.

Some skeptics also point to the fact that previous eras exhibited high enough temperatures to melt significant volumes of ice as proof that our present warming is just one of these natural trends. Various studies certainly reveal higher temperatures during certain eras such as the Medieval Warm Period, but average temperatures then are still lower than temperatures today. Temperature measurements reveal an increasing trend since 1900 with no indications of leveling off.

The most extensive temperature data—obtained by a variety of methods—reveal what is known as the hockey stick curve. These data show relatively constant temperatures from 1000 to 1900 CE (the handle of the hockey stick), a modest rise in temperatures from 1900 to 1950 CE (the bend of the hockey stick), and a dramatic rise in temperatures in recent years (the blade of the hockey stick). Data over the past 10 years verify the rising trend with no indications of changing.

A final view of many skeptics acknowledges the presence of climate change but contends that any efforts to combat these effects are not cost-effective. This argument certainly has merit if we only consider the cost of transforming the economy to one based on sustainable energy sources. Estimates place the net cost of transitioning to renewable energy in the US at $5–7 billion per year for up to 30 years.

These costs to transition to renewable energy are significant, but failure to act jeopardizes a sustainable future for those that follow us and creates a large economic toll due to increased natural disasters. Assuming four additional natural disasters per year with a cost of $10 billion each, totals $40 billion per year or $200 billion over 5 years. During a 20-year span the cost to a country like the US is $800 billion (or about 3% of the US GDP in 2023). In most other countries these impacts are greater given their smaller economies.

Beyond the economic impacts, the survival of many species on Earth is in jeopardy. In my view, we have no choice. Human activities have created problems, and we must respond to fix them. In principle, human-caused global warming will stop once we reach a state of carbon neutrality.

SOLUTIONS TO CLIMATE CHANGE

To mitigate (or reverse) global warming requires novel solutions and collective efforts by citizens. Given the magnitude of this monumental task, a multi-faceted approach seems like our best option. Individual contributions include incorporating conservation into our daily lives and advocating for utilization of alternative energy sources. Government incentives further encourage transition to renewable energy sources. Businesses and corporations also contribute by adopting sustainable practices and investing in more energy-efficient equipment.

Energy conservation does not require special devices or training but rather abilities to recognize energy-saving strategies and a desire to implement them. For reference, if 100 million families reduce 80 W of lighting by 1,000 h per year, the energy savings is 8.0×10^9 kWh. (This amount is equivalent to the energy produced by one thousand 0.9 MW power plants operating for 8,900 h.) In general, conservation efforts reduce the amount of energy required and thereby lessen the amount of CO_2 generated by reducing how much energy power plants must produce.

To determine impacts of various activities on climate change, experts evaluate carbon footprint, defined as the total amount of greenhouse gases like carbon dioxide and methane generated by a person, product, or industry. Solar, wind, hydroelectric, and nuclear energy do not release any CO_2 during the process of generating electricity; however, manufacturing parts for these industries often requires processes that release greenhouse gases. Many of these parts last for many years, thus reducing their impact when pro-rated over several years. Transition to renewable energy sources greatly reduces CO_2 emissions in the energy-producing sector. Despite their known advantages, some of these energy sources have drawbacks. For example, production of solar panels requires toxic chemicals, and hydroelectric power affects the flow of rivers and therefore impacts river ecosystems.

Determining the feasibility of replacing fossil fuels with renewable energy sources requires assessment of current needs. Worldwide power consumption (by humans) is 15 TW (15×10^{12} W) while global wind power is rated at ten times that level. Accessible wind power is estimated at 20 TW (approximately 33% greater than current requirements), so wind power itself is a viable option to replace fossil fuels.

Solar power is even more plentiful. The Earth receives approximately 10,000 times more solar power at its surface than needed by humans, thus this source has abundant potential for replacing fossil fuels. As a caveat, wind and solar sources are not always available, so adequate energy storage is necessary to fully convert to these renewables. Another alternative is to increase the number of nuclear power plants. In the US, developing and implementing a plan for long-term storage of nuclear wastes is essential to make this alternative feasible. (See Chapter 6 for more discussion.)

Identifying new energy sources that absorb CO_2 plays a role in developing new energy strategies, too. Certain microbe-based biofuels release CO_2 during combustion but absorb the gas while they grow. Under optimal conditions these two balance each other to make the process "carbon neutral." Here, the amount of CO_2 absorbed during plant growth compensates for the amount released during combustion. Recent research suggests that certain biofuels absorb more CO_2 than released during combustion, thus helping to reduce the amount of CO_2 in the atmosphere. (Algae-based biofuels have shown promise in tests involving commercial airplanes and ships.)

Other proposals to solve global warming often focus on modifying the Earth's climate to reduce atmospheric and surface temperatures. Some ideas here include seeding the atmosphere with aerosols to reflect incident sunlight or deploying large sunshades near the Earth to effectively reduce sunlight. While these innovations may help reduce temperatures, they have several drawbacks. Overall, these solutions do not reduce the amount of CO_2 in the atmosphere (or acidification of oceans) and possibly affect other enterprises such as farming and electricity production from solar arrays. Proposed geoengineering solutions also require regular maintenance to remain effective. If problems occur with maintenance, temperatures immediately increase. Implementation of these devices to manipulate the amount of sunlight reaching the Earth is difficult to model, thus their effects on climate over time are unknown.

DOES CLIMATE CHANGE LEAD TO OTHER UNEXPECTED REPERCUSSIONS?

For clarity climate change is not the definitive cause of any single weather event. However, rising temperatures tend to increase the likelihood of certain weather events like hurricanes due to higher water temperatures, particularly in places like the Gulf of Mexico and the Atlantic Ocean. Once hurricanes (and tropical storms) form, they often spawn other weather events far away from the original storm.

The case of Hurricane Beryl in July 2024 illustrates these effects clearly. Hurricane Beryl made landfall on July 8 on the Gulf Coast of Texas. Over the next several days, remnants of the hurricane moved northeastward across the US. Increased moisture and energy within this system spawned numerous tornado warnings (and touchdowns) in Texas, Louisiana, and Arkansas and heavy rains, covering a broad expanse of the eastern US. The irony here is that July is not usually a month with a high incidence of tornadoes.

HOW MUCH TIME DO WE HAVE TO RESPOND TO CLIMATE CHANGE?

This answer is uncertain; however, all mainstream scientists agree about the effects of rising CO_2 levels within our atmosphere. Surface and water temperatures will continue to increase, thereby inducing regional climate change, sea levels will keep rising, and oceans will become increasingly acidic. These effects and others will negatively impact ecosystems and make life increasingly difficult for humans. The World Health Organization (WHO) estimates that 3.6 billion people live in areas that are highly susceptible to climate change. Every effort should be made now to combat global warming and ultimately to reverse it. Successful efforts provide strategies to solve other problems such as water scarcity and emerging diseases.

SHOULD ENTERTAINERS PAY CARBON TAXES ON THEIR AIR TRAVEL?

Requiring entertainers to pay a carbon tax seems like a drastic measure to combat climate change. However, social pressures and personal convictions often are sufficient to motivate those in the public eye to compensate for their strenuous travel schedules. In early 2024, the entertainer, Taylor Swift, was widely criticized for her tour travel, along with personal travel to attend NFL games. A spokesperson for Swift reported that she purchased more than double the amount of carbon credits needed to offset all her tour travel.

While business travel is unlikely to disappear, the COVID-19 pandemic proved that many segments of our economy continue to thrive even when folks are in separate locations. All corporations, businesses, and government agencies must examine their practices to determine how much business travel is necessary and to implement schedules that balance in-person and remote work.

HOW CAN POLITICIANS WORK TOGETHER TO INITIATE EFFECTIVE CHANGE?

In the US, political solutions to climate change require bipartisan efforts in which care for the environment stimulates economic development. Climate change initiatives (like transitioning to alternative energy systems) act as drivers for the economy by creating jobs in manufacturing, installation and maintenance, sales, and service. Transforming our energy economy requires generating demand for cleaner forms of energy. In addition to building a more sustainable future, incentives are important for encouraging installation of solar panels and other renewable energy sources. With incentives, demand for sustainable energy grows, resulting in a larger fraction of our electricity coming from clean sources. As we incorporate more and more clean energy, greenhouse gas emissions and their harmful effects diminish.

Some energy producers oppose development of alternatives due to impacts on their business models. While these concerns are valid, diversification and development of other revenue streams provide certain advantages for long-term sustainability of these energy companies. Appropriate incentives (such as ones related to taxes) also provide sufficient encouragement and motivation for businesses and traditional power companies to implement renewable energy sources.

WHAT CAN INDIVIDUALS DO TO RESPOND TO CLIMATE CHANGE?

Responses to climate change vary from person to person: some folks choose to install solar panels on their homes while others devote their energies to conservation efforts. Rather than establish universal recommendations, a more productive approach calls for actions based on individual lifestyles and inclinations. To develop personal responses, several questions are worthy of consideration:

DISCUSSION QUESTIONS

1. What are my primary contributions to CO_2 production and what actions will I take to decrease these contributions?
2. What sustainable practices will I implement?
3. How will I encourage others to act?

CHAPTER 3

Human Disease and Health

HISTORICAL CONTEXT

Disease impacts every part of human existence and remains one of our enduring challenges. Old Testament writers often attributed disease to God's punishment for sin. The book of Deuteronomy commands the Israelites to put their human waste outside their camps in holes and cover them, presumably to prevent the spread of disease. Several biblical texts refer to diseases, including leprosy, pestilence, consumption, fever, and insanity, thus indicating that disease and sickness were commonplace.

By the 500s CE the first bubonic plague known as the Plague of Justinian decimated the entire Mediterranean Basin, Europe, and the near East from 541 to 549. Justinian, the Byzantine Emperor and the namesake of the plague, contracted the disease but later recovered. Estimates suggest that 15–100 million people died from this persistent and highly contagious disease during its two-century recurrence.In the 14th century, a second bubonic plague pandemic (Black Death) ravaged most of Europe from 1346 to 1353. As many as 50 million people perished during those years, thus greatly impacting religious, social, and economic norms of the day.

During World War I, the 1918–1920 influenza pandemic (also known as the Great Influenza Epidemic) infected a third of the global population

DOI: 10.1201/9781003511472-3

in four successive waves. The first recorded case of the influenza occurred in the state of Kansas (United States) in March 1918. By April officials in France, Germany, and the United Kingdom reported other cases. The H1N1 subtype of the influenza A virus caused this worldwide outbreak and later initiated the 1977 Russian flu and the 2009 Swine flu pandemics.

In 1981, the HIV/AIDS began and continues to this day. By 2023, total HIV/AIDS deaths reached 40.4 million and global cases totaled 39 million. According to the 2015 Global Burden of Disease Study, cases of the disease peaked in 1997 at 3.3 million per year. From 1995 to 2005, global cases declined to a yearly total of 2.6 million. Today, total new infections are 1–1.3 million per year. As a disease, HIV originated in nonhuman primates and then jumped to humans within the past 200 years. Initially, AIDS resulted in certain death within a few months or years, but antiviral medications have extended the lives of many of those infected.

The COVID-19 pandemic began in late 2019 and accounted for 20 million deaths worldwide. Many hospitals overflowed with patients, the supply of ventilators dwindled to the point that makeshift ones were used, and most schools instituted virtual learning for students. As a result of the pandemic, US life expectancy fell to 77.3 years—the lowest since 2003 and about 1.5 years less than in 2019.

Other pandemics have occurred during human history. Some of the more prominent ones are a third bubonic plague (1855–1960) that claimed 12–15 million lives, the 1957–1958 influenza pandemic, and the Hong Kong flu (1968–1969).

Today many of the diseases that caused previous pandemics are nearly nonexistent due to modern medical treatments and the availability of vaccines that prevent these diseases. Despite much progress, transmissible diseases such as coronavirus still threaten the human population and continue to cause substantial infections when vaccines are not available, when people refuse vaccines, or when people do not take precautions.

In the absence of epidemics and pandemics, the leading causes of US adult deaths are generally noninfectious diseases such as heart disease and cancer. In 2022, 702,880 Americans died from heart disease, accounting for about 29% of all deaths in the country. Other leading causes of death in the United States include cancer, unintentional injuries, COVID-19, stroke, chronic lower respiratory diseases, Alzheimer's disease, and diabetes.

DISEASE

Disease refers to any condition that diminishes normal body functions. Infectious diseases have numerous origins, including bacteria, fungi, protozoa, multicellular organisms, and aberrant proteins known as prions. By contrast, noninfectious ones constitute all other known diseases, including most kinds of cancer, heart disease, and genetic disease.

One of the most devastating diseases of the past half-century is coronavirus 2019 (COVID-19), a viral disease caused by the coronavirus SARS-CoV-2. COVID-19 is highly contagious with a broad range of symptoms, including fatigue, cough, fever, difficulty breathing, and loss of smell and/or taste. Transmission of COVID-19 occurs primarily by breathing droplets expelled from the lungs of infected people. The virus is quite insidious given that over 30% of those infected experience no symptoms and therefore take no precautions. The virus often attacks the respiratory system, thus causing pneumonia and then secondary infections.

While the pandemic claimed many lives and showed the gullibility of some people to false claims and conspiracy theories, one of the great stories of the pandemic was the development of a COVID-19 mRNA vaccine within a year after the initial outbreak. Unlike traditional vaccines, mRNA vaccines do not introduce a weakened form of the virus into the body. Instead they use a piece of mRNA generated in laboratories to train our cells to produce a protein or part of one that triggers an immune response upon exposure to the actual virus. The piece of mRNA introduced into the body corresponds to a particular viral protein, typically one on the outer membrane of a particular virus. Introduction of mRNA produces some of the viral proteins that the body recognizes as foreign. The body responds by creating antibodies to fight the rogue protein. Contrary to some claims, mRNA vaccines do not change DNA, nor do they enter the cell nucleus. Treatment for coronavirus focuses on managing the symptoms and ensuring that patients have adequate lung function. Antiviral medications are effective if administered within a few days of developing symptoms.

As the leading cause of adult deaths in the United States, heart disease is prevalent in many adults due to both hereditary factors and lifestyle choices. Heart disease generally includes coronary artery disease, high blood pressure, and cardiac arrest. Heart disease risks include high blood pressure, high cholesterol, smoking, obesity, unhealthful diet, physical inactivity, and excessive alcohol use.

Coronary artery disease (atherosclerosis) impacts the heart's major arteries, primarily due to the buildup of plaque that mostly consists of cholesterol, fat, calcium, and blood cells. The most imminent danger of atherosclerosis is reduction of oxygen-rich blood flow to the body. Other complications may occur if plaque buildup dislodges from the arterial walls, leading to heart attacks and strokes. Symptoms of hardening of the arteries vary, but some of the more prominent ones are pressure, tiredness, loss of appetite, nausea, swelling in the hands and feet, changes in urination, and itchiness or numbness.

Treatment of coronary artery disease ranges from changes in lifestyle to major surgery. A less noninvasive technique known as angioplasty helps increase blood flow by placing a stent inside the arteries in locations where buildup is present. During the procedure, a specialist inserts a catheter into a blood vessel in the groin or wrist and directs it to the site of the blockage. Once the catheter is secure, a balloon-type device opens the artery, and the surgeon places a stent in the location of the blockage to increase blood flow. The stent is a hollow, mesh device that presses plaque against the inside of the artery and expands the artery to allow more blood flow.

Another condition associated with heart disease is high blood pressure, often referred to as hypertension. When hypertension occurs, pressures against the artery walls and other parts of the cardiovascular system are too high (generally defined as exceeding a blood pressure of 140/90). Hypertensive people often experience no symptoms but still develop heart disease and stroke, especially if their condition remains untreated. In mild cases of hypertension, patients practice self-care by changing their diets and exercising more regularly. If high blood pressure persists, patients often take medication. Examples of medications that relax blood vessels to lower blood pressure include angiotensin-converting enzyme (ACE) inhibitors and calcium channel blockers.

Cardiac arrest refers to a sudden, unexpected loss of heart function, breathing, and consciousness due to an electrical disturbance in the heart. An arrhythmia, known as ventricular fibrillation in which the heart quivers, often accompanies cardiac arrest. During cardiac arrest, blood does not flow effectively, and the body becomes starved of oxygen, resulting in the sudden symptoms described. At the onset of cardiac arrest, the patient needs CPR and/or administration of a defibrillator to restore the heart's regular rhythm and normal pumping action. A survivor of cardiac arrest usually requires medication to treat the arrhythmia and possibly an implantable device to maintain heart rhythm.

One of the most effective methods of diagnosing hardening (calcification) of arteries is computed tomography (CT), which sends X-rays through the body to create detailed images of internal organs, bones, blood vessels, and soft tissues. CT images typically show cross-sections of parts of the body, thus providing more clarity when examining blood vessels. Assessment of these images yields calcium scores, which indicate the degree of arterial plaque buildup. When arteries have reductions in diameter of 10% due to plaque buildup, blood flow decreases by about 30%.

Calcification tests are particularly useful for those with a family history of heart disease, people with high blood pressure or high cholesterol, smokers, people with sedentary lifestyles, people over 55, people who are diabetic or overweight, and postmenopausal women. Screening for calcification of arteries is now available at most local hospitals. As a complement to specialized CT scans to examine arteries, radiologists now analyze standard mammograms to look for calcification of arteries in the breasts. The presence of calcification there likely indicates calcification elsewhere and the possibility of heart disease.

Cancer is a disease in which some of the cells in the body grow uncontrollably and spread to other parts of the body. Cancer cells differ from normal cells in that they grow even when they do not receive signals to do so, they ignore signals that usually tell cells to stop dividing, they invade other parts of the body, they instruct blood vessels to grow toward tumors, and they rely on different kinds of nutrients than normal cells. While these characteristics provide desirable conditions for cancer cells to grow (and survive), they also offer strategies for cancer treatments. For example, certain therapies prevent blood vessels from growing toward tumors, thus starving them of oxygen and nutrients.

As indicated cancer manifests as uncontrolled cell division, but its fundamental cause arises from changes to genes within our bodies. A gene is a basic unit for passing along hereditary information. Genes are specific sequences of DNA; they code for certain traits. Humans have 23 pairs of chromosomes (or 46 total) in each cell. The human X chromosome contains about 900 genes while the Y chromosome has only 55 genes. In general, biological females have two X chromosomes while biological males have one X and one Y chromosome.

Genetic changes (abnormalities) arise due to several effects such as errors that occur during cell division, damage to DNA due to harmful substances, and inherited genes from parents. The body often eliminates cells containing damaged DNA by natural processes. However, these mechanisms of removal deteriorate in the human body with age, thus aging people develop cancer with greater frequency.

Cancer spreads from one part of the body to another by a process known as metastasis. When new cancer cells form in another part of the body, they resemble the original cancer cells. Metastatic cells also exhibit similar molecular features such as chromosomal changes that are akin to their predecessors. Metastatic tumors often deleteriously affect how the body functions and often lead to death of the victim.

Experts generally classify cancer according to where it originates in the body or in what kinds of cells. The most common type of cancer is carcinoma, formed by epithelial cells that cover the inside and outside surfaces of the body. Carcinomas account for 80%–90% of all cancer cases. Most breast, skin, and kidney cancers are examples of carcinomas. Another common form of cancer is sarcoma that develops in bones, soft tissues, and connective tissues. A liposarcoma, for example, is a fat tissue cancer that typically starts in the thigh or abdomen.

Leukemias begin in the blood-forming tissue of bone marrow but do not form tumors. Instead, leukemias cause the build-up of large numbers of white blood cells that hinder the body's ability to fight infection.

Lymphoma is a type of cancer associated with the body's disease-fighting network (lymphatic system), including the lymph nodes, spleen, thymus gland, and bone marrow. Common symptoms of lymphoma include enlarged lymph nodes, fatigue, and weight loss.

Multiple myeloma is a cancer of plasma cells. When multiple myeloma occurs, plasma cells become cancerous and multiply. This kind of disease weakens bones, thus leading to a greater incidence of bone fractures. Most people diagnosed with stage I or II multiple myeloma are still living 5 years after diagnosis.

Physicians screen for cancer by several methods, including routine physical examinations and review of a patient's family history. Most wellness examinations also include analysis of blood to check for levels of certain chemicals that may indicate cancer. Imaging tests such as those obtained by CT, MRI, ultrasound, X-rays, and PET are extremely useful for looking at internal structures, including organs, to see if cancerous tumors are present.

In certain cases, doctors use a specialized tool known as an endoscope, equipped with a light and camera, to look for tumors or other abnormalities inside the body. When certain tissues look suspicious, physicians remove portions of the tissue and examine them for cancer. These tests known as biopsies also indicate the type of cancer and its level of aggressiveness.

CANCER TREATMENTS

Chemotherapy incorporates drugs that kill cancer cells by disrupting the cell cycle. One class of chemotherapy drugs, known as alkylating agents, include cisplatin and cyclophosphamide. These agents replace hydrogen atoms on DNA with alkyl groups, thus creating cross-links in the DNA chain. Programmed cell death follows due to the inability of these cells to synthesize DNA, RNA, and proteins. Common uses of these agents are for treatments of leukemia, lymphoma, breast, lung, and ovarian cancer. While chemotherapy is highly effective, several side effects, including mouth sores, nausea, hair loss, and fatigue, are common. Patients often receive steroids in conjunction with chemotherapy treatments to alleviate some of these effects.

Hormone therapy uses specialized drugs to block or reduce hormone levels in the body to slow or stop the growth of cancer cells. This kind of therapy is effective for treatment of hormone-induced cancers such as prostate and breast cancers. In the case of breast cancer, the therapy blocks the growth of hormone-receptor-positive breast cancer. Women receiving hormone therapy for breast cancer might experience side effects such as hot flashes, vaginal discharge, dryness or irritation, fatigue, and nausea.

Hyperthermia refers to a cancer treatment that utilizesthermal energy to damage and/or kill cancer cells. Hyperthermia is effective in treating tumors close to the surface of the body. Several methods for delivering energy to tumors are available. One of them consists of a needle-like probe that delivers radiofrequency energy once the probe is inserted into the tumor.

Immunotherapy takes advantage of the body's natural immune system to fight cancer. Immunotherapy enters the body through an IV and alleviates cancer by helping the immune system work harder or in more targeted ways. Immunotherapy works by slowing the growth of cancer cells, preventing cancer from spreading, and boosting the immune system to rid the body of cancer cells.

Photodynamic therapy works by injection of specialized medicine, activated by electromagnetic waves (light) to form substances that kill cancer cells. The injected medicine remains inside cancer cells longer than normal cells. Light directed toward those areas that retain the medicine activates it, thereby killing cancer cells selectively.

Cryotherapy refers to using cold substances (such as liquid nitrogen or gas from it) to freeze external cancer cells and destroy them. Cryotherapy is particularly effective for skin cancer and precancerous spots or lesions. When needed, specialized instruments are available to deliver cryotherapy to tumors inside the body, too.

Radiation therapy destroys and/or inhibits the growth of cancer cells through doses of ionizing radiation, such as X-rays, gamma rays, neutrons, protons, or other sources. In addition, the use of energetic calcium ions to treat cancerous tumors is an emerging technology. External-beam radiation therapy directs particles or waves to tumors within the body. Internal radiation therapy or brachytherapy treats cancer by delivering radioactive material via implants to cancerous regions. The proximity of the material ensures that the cancerous cells receive a higher dose of radiation.

Systemic radiotherapy employs radioactive drugs, administered by swallowing or injection to target cancer cells. Once inside the body the radioactive substance accumulates near tumors to selectively destroy cancer cells there. This therapy is particularly effective for treating bone, prostate, and thyroid cancer. Here the use of radioactive materials often causes side effects such as fatigue, nausea, vomiting, and diarrhea.

HOW DOES CLIMATE CHANGE CONTRIBUTE TO DISEASE AND POOR HEALTH?

Impacts of climate change on health are far-reaching and not entirely known. Exposure to higher extreme temperatures obviously causes extra stress on the body and exacerbates conditions such as cardiovascular disease, diabetes, asthma, and mental health problems. When stressed the body often experiences various heat-related illnesses such as heat cramps, heat exhaustion, and heat stroke. Heat stroke results in elevated body temperature with accompanying damage to the heart, brain, and kidneys. Heat exhaustion generally occurs due to dehydration, resulting in symptoms such as nausea, tiredness, and heavy sweating. Heat cramps usually manifest as painful muscle spasms caused by dehydration and loss of electrolytes.

Extreme weather also threatens water supplies and disrupts electrical service. Lack of electricity during and after severe storms causes severe strain on many people—especially those who need air conditioning and specialized medical equipment. Rising temperatures negatively impact the water cycle, leading to unpredictable rainfall patterns and rising sea levels. As a result, about half the world's population experiences water scarcity. Climate change also impacts water quality due to increases of contaminants during downpours and algal blooms in warmer waters.

Climate change affects air quality in several ways. Higher temperatures generate more ground-level ozone (and thus less breathable oxygen) due to increased reactions that produce ozone. Smog and smoke from wildfires also increase, leading to more difficulty breathing, allergy flare-ups, and asthma.

The threat of significant storms with much longer seasons also creates anxiety and contributes to several of the mental health conditions described below. The upshot is that many of us who live in impacted areas must adapt to live with more uncertainty and prepare for a variety of contingencies. Populations that are most vulnerable to rising temperatures and overall climate change include outdoor workers, older adults, low-income households, and people with certain medical conditions.

WHAT IS MENTAL HEALTH AND HOW DO WE MAINTAIN SOUND MENTAL HEALTH?

Mental health generally refers to an individual's emotional, psychological, and social well-being. The state of a person's mental health determines their thought processes and how they feel and act. For example, a depressed person often cannot think clearly enough to monitor their eating habits and water intake, so their bodily health suffers due to their mental state. Mental health also impacts how well people handle stress, interact with others, learn, and contribute to their community.

Mental health conditions vary but range from depression and anxiety to addiction and phobias. Factors that determine our state of mental health include restorative sleep, physical activity, healthful eating, social connections and interactions, and opportunities to help others. Treatments for mental health disorders range from self-care to medication to various kinds of therapy. To maintain sound mental health, experts encourage people to exercise regularly, eat healthfully, sleep regularly, remain connected with others, practice gratitude, find outlets in nature, avoid harmful substances, and remain creative.

WHAT ARE THE ESSENTIALS FOR MAINTAINING GOOD HEALTH?

The most comprehensive approach for maintaining vigorous health combines a balanced diet and plenty of exercise with elimination of poor habits. A healthful diet consists of various food groups, containing sufficient calories for the person's age and lifestyle. Experts recommend that each main meal consist of a high-fiber food, typically a starchy carbohydrate such as potatoes, bread, rice, pasta, and cereals. For reference, one serving of cooked white rice has about 53 g of total carbohydrates. Carbohydrates contain about half the calories of fats per gram, so limiting fat intake is prudent, too. One effective way to reduce fats is to limit the amount of butter and dressings added to various dishes.

Along with high-fiber carbohydrates, most physicians and dietitians recommend five servings of fruits and vegetables daily. A serving of vegetables is about 75–80 g, corresponding to about one cup of raw leafy ones. Fish is a healthful food that contributes protein, vitamins, and minerals to our diets. Experts also recommend two servings of fish per week, including one serving of the oily variety. Oily fish are rich in omega-3 fatty acids and include salmon, trout, herring, sardines, pilchards, and mackerel.

To improve cardiovascular health, aerobic exercise with sufficient intensity is necessary. Higher intensity workouts correspond to higher target heart rates. Target heart rates depend on both the resting heart rate RHR of the individual and another quantity known as heart rate reserve HRR. Heart rate reserve is the difference between the individual's maximum heart rate and their resting heart rate.

Statistical data show that maximum heart rate is approximately 220 minus a person's age. Taking a percentage (called percentage training intensity, %TI) of the heart rate reserve and adding that value to the resting heart rate determines the target heart rate. For reference, a 0%TI indicates the person's heart rate during the workout is their resting rate, while a 100%TI indicates their heart rate during the workout is their maximum heart rate. To promote aerobic health, experts recommend TIs in the range of 50–80%. As an example, consider a 20-year-old with a resting heart rate of 60 bpm who desires to train at 70%TI. From the considerations above, their target heart rate is (0.7)(220-20-60) + 60 or 158 bpm. High training intensities increase cardiovascular efficiencies and lung capacity.

Vigorous exercise not only improves cardiovascular health but stokes the body's metabolism so that the body consumes excess energy for several hours after the activity. Numerous studies indicate that 45 min or more of vigorous exercise—defined as the body consuming oxygen at 75% of its maximum consumption volume, VO_{2max}, or more—results in expenditure of approximately 200 extra food calories. For reference, VO_{2max} refers to the maximum amount of oxygen a person consumes and typically depends on fitness levels. The presence of this "afterburn effect" has implications for designing diets, managing weight loss, and recovering after exercise.

Another aspect of maintaining health is to eliminate unhealthful habits such as overeating and tobacco use. While obesity continues to increase in the US, tobacco use among adults is at an all-time low. Data from 2022 indicate just under 20% of US adults report using tobacco products.

SUMMARIZE THE COVID-19 PANDEMIC RESPONSE IN THE US

In the US the number of COVID-19 cases exceeded 111 million, resulting in approximately 1.2 million deaths. Many of these deaths were preventable if people would have followed basic health guidelines. However, once attitudes toward the pandemic were influenced by political rhetoric, many US citizens refused to wear masks, continued to gather in large groups, and declined vaccinations.

Interviews confirmed that people often embraced misinformation and made poor decisions based on false claims. Several people who contracted COVID-19 believed the pandemic was a hoax, as purported by several conspiracy theorists. For some the realization that the disease is real and deadly came too late. The grim news is that another global pandemic would likely claim even more US lives, given the fatigue with following protocols and safe practices. Other parts of the world handled the crisis much more effectively and experienced far lower morbidity rates. For example, more isolated countries like New Zealand, which implemented strict lockdown policies, experienced few deaths.

CHAPTER 4

The Chemical Medley

COMBINATIONS OF PROTONS, NEUTRONS, and electrons constitute every known element in the universe. These units of matter serve as the building blocks of all molecules, ranging from simple diatomic (H_2, O_2, and N_2) ones to long-chained polymers composed of multiple repeating units. Many chemical reactions such as oxidation of metals in the Earth's atmosphere occur seamlessly under ordinary conditions. For example, iron (Fe) readily reacts with atmospheric oxygen in the presence of ambient moisture to produce iron (II) oxide and iron (III) oxide, commonly known as rust.

Other reactions require either catalysts and/or certain conditions of pressure and temperature to proceed. Raw petroleum consists of high-molecular-weight hydrocarbons that must be broken down into smaller molecules for use as fuels. Zeolite catalysts, in the presence of elevated temperatures, facilitate this process in reactions that produce petroleum coke and liquid fuel that serves as the starting material for gasoline. In the presence of sunlight and chlorophyll, plants convert carbon dioxide and water into carbohydrates via photosynthesis. Once formed, these carbohydrates undergo a series of reactions to generate a vast array of compounds that comprise plant matter. Interactions of molecular hydrogen (H_2) with metal catalysts such as palladium (Pd) metal at elevated temperatures promote dissociation into H atoms and migration of H atoms

DOI: 10.1201/9781003511472-4

into spaces (interstices) within the metal lattice. Hydrogen stored in this fashion is useful in a hydrogen economy.

Chemical processes are vital for producing desired products, carrying information, and storing and/or absorbing energy. The sections below review some of the seminal contributions of chemistry, discuss water as a chemical and natural resource, and provide a summary of organic molecules and their uses and applications.

CHEMICAL BREAKTHROUGHS

The development of electric cars along with other consumer-electronic devices including cell phones and laptop computers was facilitated by the invention and commercialization of the Li-ion battery. The ability of Li ions to reversibly intercalate within electronically conducting materials is the foundation of rechargeable Li-ion batteries. One of the workhorses of handheld electronics is the lithium polymer battery, consisting of a polymer gel as the electrolyte, a graphite anode, and a lithium cobalt oxide cathode. During discharge Li ions move from the negative electrode (anode) to the positive one (cathode) through the electrolyte and separator diaphragm. At the same time, an electron current develops in an external circuit, with electrons flowing from the anode to the cathode. The presence of current in an external circuit provides electrical power to operate devices.

The widespread application of lithium-ion batteries has led to the e-revolution in which individuals have acquired the ability to own powerful personal electronics. For their work in developing the lithium-ion battery Akira Yoshino, M. Stanley Whittingham, and John B. Goodenough received the 2019 Nobel Prize in Chemistry.

One of the great stories of humanity responding to an environmental crisis occurred in 1987 with the signing of an international agreement to phase out production and use of harmful chlorofluorocarbons (CFCs)—chemicals that contribute to depletion of ozone in the atmosphere. Ozone (O_3) in the Earth's upper atmosphere provides natural protection from ultraviolet (UV) radiation by effectively absorbing photons in this region of the electromagnetic spectrum. UV energy from the Sun has harmful effects on humans, primarily due to increased skin cancer and prevalence of cataracts in the eye. More generally, UV waves damage DNA of plants and animals, inhibit plant growth, and reduce photosynthesis in plankton.

Depletion of atmospheric ozone occurs via a catalytic reaction, initiated by introduction of harmful molecules containing chlorine and/or bromine into the atmosphere. Prior to the late 1980s the primary culprits of ozone depletion were chlorofluorocarbons—compounds used as propellants in aerosol sprays and coolants in air conditioning. Once in the atmosphere these molecules naturally rise to the stratosphere where UV radiation initiates their breakdown to form chlorine atoms. Chlorine atoms then react with ozone (O_3) to produce chlorine monoxide and oxygen gas (O_2). The chlorine monoxide subsequently reacts with a free oxygen (O) atom to regenerate the chlorine atom and O_2. Once freed, the chlorine atom reacts with another O_3 molecule, thus one chlorine atom destroys multiple ozone molecules. In polar regions (such as over Antarctica) "polar vortexes" occur during winter months, thereby trapping chlorine compounds and accelerating ozone depletion. Beginning in 1989 the use of CFCs diminished significantly due to implementation of an international treaty known as the Montreal Protocol.

Since humankind first developed tools, production of new materials has been paramount to our continued advancement. Moreover, the ability to carefully control material properties continues as an emerging area of research and development. Quantum dots (QDs) are a class of materials—composed of particles with characteristic sizes of a few nanometers—whose electrical and optical properties depend critically on those dimensions. QDs restrict electrons to well-defined spaces, thus electrons behave as quantum-mechanical particles in boxes and thus occupy discrete energy levels. The presence of well-defined electronic energy levels assures absorption and emission of electromagnetic (EM) radiation at specified frequencies (wavelengths).

The correlation between size and material properties allows researchers to finely tune QDs for specific applications. Potential applications of QDs focus on development of novel materials for single-electron transistors, single-photon generation, and quantum computing.

WATER AND THE HYDROSPHERE

Water (H_2O), as a molecule, contains two hydrogen (H) atoms joined to an oxygen (O) atom by electron sharing (two covalent bonds). Once formed the oxygen atom has two lone pairs of electrons. Given this structure additional negative charge resides near the oxygen atom, thus leaving the hydrogen atoms more positively charged. This separation of charge gives water its polar nature and leads to effective hydrogen bonding between water molecules. Hydrogen bonding occurs through the lone pairs of oxygen atoms whereby each oxygen atom forms two hydrogen bonds with nearby hydrogen atoms. Liquid water remains intact due to the formation of multiple hydrogen bonds around each molecule.

When ice forms each oxygen atom binds to two H atoms via hydrogen bonding. Moreover, each H atom hydrogen bonds to an O atom. The combination of the ordinary bonds in H_2O and hydrogen bonding produces a lattice in which each O atom connects to four other O atoms via hydrogen atoms, thus leading to a honey-comb structure with large open spaces. Overall, a given number of atoms in ice occupies more space than the same number in liquid form, thus, upon freezing, water expands and has lower density. (Water's maximum density occurs at 4°C.) Such effects are observed when a bottle of water breaks upon freezing and when ice floats on a body of water. Most solids increase in density when they transition from liquid to solid, so they sink in their own liquid.

Changes in water density contribute to the turnover of water within bodies such as lakes. During transition from warm to cold temperatures, warm surface water sinks as it cools and becomes denser. During transition from cold to warm temperatures, cold surface water near the freezing point of water sinks as it warms and becomes denser. When ambient (air) temperatures are constant, water turnover is minimal.

The fact that water ice floats in its own liquid allows many organisms to flourish underneath a layer of ice on a partially frozen lake. If water behaved like other liquids, it would freeze from the bottom, leading to vastly different kinds of aquatic life. Moreover, freezing from the bottom would have prevented the development of the rich variety of marine life that we see today.

Ice also contributes to weathering of rocks and formation of potholes when liquid water freezes and expands within cracks. Freezing of water in the ground leads to a phenomenon known as frost heave whereby formation of ice in the soil produces an upward (or outward) movement of the ground surface. Frost heave has the potential to lift structures embedded in the ground, so contractors bury concrete columns to depths below the frost line to prevent damage to buildings and other structures.

Water has several other unique properties such as high values for freezing point, boiling point, heat of vaporization, and heat of fusion, relative to other molecules of the form H_2X. Higher freezing and boiling points indicate that higher temperatures are necessary to cause transitions from the solid to liquid and liquid to vapor states. Moreover, the higher heats of vaporization and fusion indicate greater energy requirements to produce these phase transitions. These properties ensure that water is stable over a wide temperature range.

Water also has suitable properties that make it indispensable within the Earth's ecosystem. Compared to nearly all other liquids, water has a higher specific heat, thermal conductivity, and surface tension. High surface tension results from strong attractive forces between molecules due to hydrogen bonding. These strong attractive forces ensure that water effectively rises in stems and roots via capillary action. High surface tension also facilitates accumulation of water molecules in small voids between particles in the soil.

High thermal conductivity translates to efficient transfer of energy via water molecules, thus moderating temperatures in places like the Earth's atmosphere and large bodies of water. Given water's high specific heat, temperatures of warm-blooded species maintain their desired range due to the ability of water to absorb or transfer energy without appreciable temperature changes. Water's high heat of vaporization also helps humans transfer large quantities of energy when water evaporates from their skin (with minimal loss of fluids). Water is a nearly universal solvent for polar covalent and ionic compounds due to its polar nature. Thus, water is an ideal transport medium for ions needed for plants and animals.

Certain chemical reactions generate water molecules; however, most water molecules undergo circulation on Earth by a continuous water cycle. Water constantly enters the Earth's atmosphere via evaporation from surface water and the soil. Plant stems and leaves also introduce water into the atmosphere via a process known as transpiration. Average residence time for water in the atmosphere is about 10 days, whereas average residence time for water in the oceans is thousands of years. In each region of the world, precipitation and evaporation depend on various factors, including temperatures, winds, and local atmospheric pressures.

Rain and other forms of precipitation carry water to the Earth naturally, thus producing a major source of clean water. As precipitation falls it carries dust, soluble particles, and some gases from the atmosphere. When certain compounds such as SO_2 are present in air hydrogen (H^+) and sulfate (SO_4^{2-}) ions form inside raindrops, leading to acid rain. (See Chapter 9 for more details.) Following precipitation, salts also dissolve in water as it flows to lakes and other bodies. Over time, bodies of water such as the Dead Sea in Israel become rich sources of salts. Natural water also contains dissolved CO_2, which in solution produces hydrogen (H^+) and hydrogen carbonate (HCO_3^-) ions. This reaction is reversible, allowing natural waters to maintain constant acidity and readily dissolve minerals. Thus, flowing natural waters have the capability to transport minerals over vast distances.

Water purification occurs in nature by several processes within the water cycle and the action of microorganisms. Solids suspended in water settle to the bottom of quiet lakes and pools; alternatively, seepage of water within soil removes many particulates, too. The flow of water over shallow rock beds removes dissolved gases and volatile impurities. Organic matter in water decomposes via aerobic decomposition to form harmless molecules and ions. If enough oxygen is available, aerobic microorganisms keep the water free of organic matter.

In the US water that drains from all parts of homes, along with industrial wastewater, requires treatment before it is available again in the water supply. The first stage of treatment (known as primary sewage treatment) filters out large pieces of material and then allows sediments to fall in large sedimentation tanks. At the end of the first stage, 40%–60% of the suspended solids and 25%–35% of the oxygen-demanding wastes are removed.

Secondary sewage treatment removes up to 90% of oxygen-demanding wastes. Here methods typically expose the sewage to a large population of aerobic bacteria and plenty of oxygen. Various schemes then promote sufficient water mixing with O_2 in the presence of appropriate bacteria. Tertiary water treatment removes additional materials, as needed.

As a resource water is indispensable to our existence and is necessary for the daily activities of nearly all animals and plants. While the water cycle continues to provide water via natural processes, the availability of water in certain locations on Earth is dangerously scarce. Current data indicate that about 2 billion people on Earth do not have access to safe drinking water. Additionally, roughly half the world's population experiences severe water scarcity for at least part of the year.

ORGANIC MOLECULES AND CHEMISTRY

The linchpin of all organic chemistry is the carbon atom with its ability to join with four atoms (or groups of atoms) via tetrahedral bonding. Nearly all organic compounds have a carbon chain with hydrogen and/ or other functional groups attached to that chain. The simplest of the hydrocarbons are the saturated ones with all covalent single bonds: each carbon atom attaches to four other atoms, and each hydrogen atom connects by one bond to a carbon atom. Methane is a saturated hydrocarbon with one carbon atom and four hydrogen atoms (CH_4) and the primary component of natural gas. Methane is part of a family of compounds called alkanes that have the form C_nH_{2n+2}. Simple saturated hydrocarbons consist of straight-chained compounds where the carbon backbone is straight. Carbon atoms also form stable shapes such as rings in molecules like cyclohexane.

In contrast with saturated hydrocarbons, unsaturated ones have covalent double or triple bonds. Hydrocarbons with covalent double bonds are called alkenes; ones with covalent triple bonds are known as alkynes. Acetylene (or ethyne) consists of two carbon atoms joined by a covalent triple bond with one hydrogen atom attached to each carbon atom. Acetylene is a gas often used in welding. When oxidized, acetylene's triple bond releases large quantities of energy capable of generating temperatures around 2,200°C.

Aromatic hydrocarbons are another class of unsaturated molecules that contain resonance-stabilized ring structures. Benzene and its derivatives are some of the most common aromatic hydrocarbons. The primary uses of benzene are in the manufacture of other chemicals to make plastics, resins, synthetic fibers, lubricants, dyes, and drugs.

Hydrocarbons are found primarily in fossil fuels—natural gas, petroleum, and coal—formed by the decay of plant and animal remains over millions of years. Petroleum taken from the ground requires various chemical processes to increase the yield of molecules suitable for combustion in vehicles and other applications. The first step in refinement is distillation and extraction of fractions with different boiling points. The lower boiling point fractions, called petroleum ether, typically consist of solvents such as pentanes and hexanes. After distillation, the leftover oils and greases are suitable for lubrication while the heavy residue serves as roofing and paving material.

Using heat and catalysis, isomerization transforms straight chain alkanes into branched ones that perform better as fuels. Another process known as petroleum cracking, breaks large molecules into smaller ones that are suitable for gasoline. Gasoline generally consists of a homogeneous mixture of lightweight hydrocarbons with 4 to 12 carbon atoms per molecule. Some of the specific compounds include isooctane and butane. Other processes performed during petroleum refinement include petroleum alkylation and petroleum reforming. Petroleum alkylation refers to combining lower-molecular-weight alkanes and alkenes to form molecules in the gasoline range. Petroleum reforming produces hydrocarbons needed for other purposes besides gasoline.

Organic compounds have far-ranging and diverse applications due to the wide array of functional groups that are available to connect to simpler hydrocarbons. These functional groups consisting of an atom or a collection of atoms contribute to the properties of molecules to which they attach, thus affecting their behavior.

One of the simplest functional groups consists of a single halogen atom that substitutes for a hydrogen atom in a saturated hydrocarbon. When this occurs, the saturated hydrocarbon becomes an alkyl halide (RX), where R is a saturated group and X is fluorine, chlorine, bromine, or iodine. Many of these compounds are quite toxic and, at one time, were integral to the production of pesticides and defoliants. Most are now illegal due to harmful effects of exposure to them.

Another common functional group is the hydroxyl or OH group. When attached to a saturated hydrocarbon (R group), the R and OH groups produce alcohol. Methyl alcohol (or methanol) isknown as wood alcohol due to its production during wood distillation. Today, chemical producers generate methanol synthetically from carbon monoxide and hydrogen gas. Methanol is a common additive in blended fuels.

Ethyl alcohol (ethanol, CH_3CH_2OH) is a component of alcoholic drinks and is a valuable solvent and reagent in organic synthesis. When producing ethanol for the spirits industry, fermentation is the preferred method. Fermentation is the process of converting starches or sugars in various grains, fruits, and other natural products into ethanol. Certain enzymes convert starches to sugars; other enzymes then convert sugars into ethanol and carbon dioxide. Phenols are akin to traditional alcohols; they consist of an aromatic ring with an OH group attached (ArOH). The active component in poison ivy, urushiol, is a phenol. When exposed to skin, poison ivy causes rash and irritation due to an allergic reaction with urushiol.

Another class of organic compounds consists of an R group connected to an alkoxy group (RO) to form an ether. An ether is represented in its general form as ROR, of which diethyl ether (CH_3OCH_3) is a common example. Ethers are relatively unreactive but are valuable as solvents for many chemical processes. Ethers tend to form explosive peroxides in the presence of oxygen; therefore, workers must follow strict procedures when handling these chemicals.

Amines are a class of organic compounds in which one or more hydrocarbon groups substitute for hydrogen atoms of the ammonia molecule (NH_3). The number of hydrogen atoms replaced determines the classification of these compounds into primary (RNH_2), secondary, and tertiary amines. Several amines, referred to as alkaloids, occur naturally in plants. Nicotine is a highly addictive alkaloid present in cigars, cigarettes, and other tobacco products. Morphine is a sedative and painkiller derived from the opium plant. Coniine is another simple toxic alkaloid found in hemlock—the plant that poisoned Socrates. Drugs such as amphetamines and methedrine are stimulants in the amine family. Within the human body, the amine known as epinephrine (or adrenalin) enters the blood stream from the adrenal glands when we experience fear or stress.

The carbonyl functional group has double bonds between a carbon and an oxygen atom, with two additional bonds available on the carbon atom. An aldehyde has the form RCHO, with a hydrogen atom and a hydrocarbon group attached to the carbonyl group. A terpene aldehyde known as citral occurs in oils present in citrus fruits and lemongrass. Vanillin is an aldehyde derived from orchids of the genus *Vanilla*. Manufacturers extract vanilla for flavoring by soaking the beans in an alcohol solution. Substitution of one of the hydrogen atoms of an aldehyde with a hydrocarbon group forms a ketone. Dimethyl ketone, often known as acetone, is a common solvent. Ketones are widely used solvents needed for manufacturing explosives, lacquers, paints, and textiles.

A carboxylic acid has a doubly bonded oxygen attached to a carbon along with an OH group on the same carbon atom. When reacted with bases, carboxylic acids form salts. Lactic acid accumulates in muscles of humans when insufficient oxygen is available during breakdown of glucose to produce energy. The body's liver and kidneys filter out lactic acid and convert it to glucose.

An ester forms when a hydrocarbon group replaces the acidic hydrogen of a carboxylic acid. Aspirin is an ester generated from acetic acid (CH_3COOH) and salicylic acid. Many esters contribute to the flavoring of fruits. For example, isoamyl acetate has the characteristic smell of a banana.

Living cells contain four major classes of organic compounds—proteins, carbohydrates, lipids, and nucleic acids—along with water and several kinds of inorganic ions. Proteins consist of amino acids whose name implies the presence of an amine group (NH_2) and a carboxylic acid group. Amino acids join with each other via formation of a peptide bond, generated during the loss of a water molecule between a carboxyl group and an amine group. Twenty amino acids serve as the building blocks of all known proteins. Proteins perform a variety of functions within the human body, ranging from regulating growth and metabolism to forming muscle tissue.

Carbohydrate compounds are sugars of various complexities. Simple sugars are monosaccharides; polymers of two to ten sugars are oligosaccharides, and polymers of more than ten sugars are polysaccharides. In general, five or six carbon atom sugars are more stable in a ring versus an open structure. Glucose ($C_6H_{12}O_6$) is a common monosaccharide with six carbon atoms arranged in a ring. Both starch and cellulose are polymers of D-glucose. Cellulose contains hundreds to thousands of glucose units in which linear molecules are organized into bundles. Starch is a primary energy source in plants and consists of two polymers of D-glucose.

Lipids are nonpolar organic compounds with large hydrocarbon sections. These compounds are nonpolymeric, insoluble in water, but highly soluble in organic solvents. Glycerides are common lipids, formed by creating an ester from a fatty acid. Glycerides include a range of animal fats and vegetable oils. Many of us use a variety of these oils, including soybean oil, linseed oil, olive oil, and peanut oil in cooking and food preparation. Glycerides in the body store energy for future use. The liver and other tissues store triglycerides, which break down to release energy when needed; moreover, the body retains excess triglycerides as fat. Steroids consisting of a four-ring basic carbon unit also fall within the classification of lipids. Cholesterol is a well-known steroid linked to hardening of the arteries and cardiovascular disease.

Nucleic acids are some of the most robust biochemical molecules, consisting of repeating units known as nucleotides. Nucleotides themselves have three parts: a heterocyclic base, a sugar, and one or more phosphate (PO_4^{3-}) groups. Nucleic acids store genetic codes via deoxyribonucleic acids (DNAs) and ribonucleic acids (RNAs) and regulate protein production. The four basic nucleotides in DNA are adenine (A), cytosine (C), guanine (G), and thymine (T). DNA has a double helix structure and is crucial to the transfer of genetic information. The ability to determine the order of the basic nucleotides is known as DNA sequencing—a technique useful for identifying variations or mutations associated with diseases.

DISCUSSION QUESTIONS

1. What steps must be taken to ensure the safety of drinking water?
2. What is the public role in regulating genome editing?
3. What protocols are necessary to prepare for chemical accidents and mitigate their impacts?
4. How do we balance progress with the potential for chemical accidents?

CHAPTER 5

The Electrical Grid

Boom or Bust

APPROXIMATELY ONE-THIRD OF WORLDWIDE primary energy consumption goes into production of electric power. Various primary sources release stored energy to generate mechanical motions needed to operate turbine-electric generator systems. For distribution a host of generating stations place electric power on a system called the grid. The grid consists of generating stations, transformers that prepare electricity for transmission and distribution, high-voltage lines for long-distance transmission, and lower-voltage lines for local distribution. The US grid originated around 1882 when Thomas Edison began production of electricity at the Pearl Street Station in lower Manhattan. Edison's plant served 59 customers while today's grid provides on-demand electricity to hundreds of millions of users nationwide.

Historically, fossil fuels such as coal and natural gas served as the primary energy sources for electricity production, but today sustainable sources supply an increasing share. In the United States, renewable energy sources (wind, solar, hydropower, biomass, and geothermal) now account for over 20% of electricity production. The influx of electrical power from renewable energy sources contributes to the goal of achieving a carbon-free electrical power sector by 2035.

DOI: 10.1201/9781003511472-5

Several countries like Ethiopia, Bhutan, Albania, and Iceland derive their electricity entirely from renewable sources. In 2021 China led global renewable electricity production with a 31% share, followed by the United States (11%), Brazil (6.4%), Canada (5.4%), and India (3.9%).

BASIC DC AND AC CIRCUITRY

Charge is a basic property of matter that contributes to how systems behave electrically and magnetically. Many fundamental particles in nature, including electrons, possess charge expressed in units of Coulombs. For reference 1 mole (6.022×10^{23}) of electrons is equivalent to 96,500 Coulombs. Numerous experiments and observations suggest that two kinds of charge (positive and negative) exist in nature. Atomic nuclei contain protons that are positively charged and neutrons that are neutral.

To understand basic electrical phenomena requires a working knowledge of the quantities: current, voltage, and resistance. Current is a measure of the flow of charge through a cross-sectional area in a unit of time. To visualize this quantity, imagine counting the number of particles passing through a cross-sectional region of a wire during each second. From the known number of particles and the charge on each, the current is the amount of charge that passes through that area during each second. The unit for current is a Coulomb per second, often referred to asan Ampere (Amp, A).

Generation of current within a circuit requires a voltage (often referred to as an electric potential difference or electric potential) to supply the needed energy. A potential difference is the potential energy difference between two points in a circuit per unit of charge. A potential energy difference in electronics is analogous to a rock on a cliff possessing energy relative to an identical rock on the ground. When the rock on the cliff falls, it tends to move toward ground level (or it seeks a lower point of potential energy). Similarly, a positive charge at a higher electric potential tends to seek a lower electric potential and thus undergo transport (motion) in a circuit. DC electric potentials are constant over time, whereas AC electric potentials oscillate like a wave. (More specifically, an AC voltage oscillates as a sine wave.) The standard unit for electric potential difference is the Volt.

The resistance of a circuit is a measure of how much the circuit opposes the flow of charge (generally for a given potential difference). Resistance in a circuit arises from collisions between moving charges and atoms within the conducting material. If charged particles in motion encounter additional atoms during a specified time interval, both the number of collisions and the electrical resistance increase. The unit for resistance is the Ohm.

For simple DC circuits in which the electric potential V is constant, current I is directly proportional to V and inversely proportional to the resistance R:

$$I = \frac{V}{R}.$$

Thus, the amount of current (charge flow rate) increases as the electric potential increases. A larger electric potential translates into more energetic charges, higher drift speeds for those charges, and greater charge flow rates. (Drift speed refers to the average speed of charges as they move from a higher electric potential to a lower one.) Current also increases as the resistance decreases. Lower resistance within a conductor effectively reduces the collision rate of free charges and thereby increases the rate of flow.

Current produced in an alternating current (AC) circuit depends on the interplay of resistance (R), inductance (L), and capacitance (C). These three quantities determine the circuit impedance; the ratio of the electric potential to the impedance then determines the current. While most circuits possess all three factors, the discussion below describes each quantity separately to show how each contributes to the overall state of the circuit.

Collisions of electrons and atoms within devices determine the resistance of loads in circuits. In so-called resistive loads, collisions transfer energy from the current-carrying particles (electrons) to atoms within the conducting materials, thus increasing the temperature of the conductors. The amount of energy per time (power) delivered to a conductor with resistance (R) is proportional to R and the square of the current.

A purely inductive load has no resistance associated with the conductor itself but produces magnetic fields due to configurations of the current-carrying wires. In most cases, inductive loads have multiple turns of wire formed into a coil, such as ones in electric motors. Current-carrying coils produce concentrated magnetic fields along their axes. (Straight current-carrying wires also generate magnetic fields, albeit weaker ones.) Changing magnetic fields occurring over given regions (areas) produce electric potentials that oppose the source electric potential. As a result, the circuit has a resistance-like quantity known as inductive reactance, opposing the flow of charge. For a circuit operating at a frequency (f), the inductive reactance (X_L) in units of Ohms is

$$X_L = 2\pi f L$$

where L is the inductance of the coil or other circuit element. The frequency of an AC circuit refers to the number of cycles (or oscillations) that occur each second.

A purely capacitive load consists of conductors arranged so that application of an oscillating voltage causes time-dependent charge variations on those conductors. As charge accumulates on these conductors (separated by insulating material), electric fields develop. When electric fields build up, they inhibit the flow of charge due to changes in energy associated with the free charges. Capacitors produce capacitive reactance, given by

$$X_C = \left(2\pi f C\right)^{-1}$$

where C is the capacitance of the circuit (or circuit element) in units of Farads.

When all three quantities (resistance, capacitive reactance, and inductive reactance) appear in a circuit, they combine mathematically to give the impedance (Z). Impedance characterizes how much the circuit opposes the flow of charge, thus a decrease in impedance results in an increase in current. The impedance for a circuit with R, L, and C is

$$Z = \sqrt{R^2 + \left(X_L - X_c\right)^2}.$$

Current in an AC circuit is the ratio of the voltage to the impedance (Z). When X_L and X_C match one another, the impedance reaches a minimum, thus producing maximum current and power. In many practical cases, adjustment of capacitive and inductive reactances is desirable to produce maximum current and power in a circuit. For example, in a home heating and cooling system, a large fan (with high inductance) forces air through the ductwork for distribution throughout the house. To maximize the current of such a circuit, designers often add a capacitor so that the inductive and capacitive reactance match one another.

OVERVIEW OF ELECTRIC GRIDS

Electric grids require generating stations that produce alternating voltage for distribution to sites near and far. Energy derived from a primary source performs work to turn a turbine-generator system that places electrical power on a grid system. An AC generator consists of a coil (of wire) rotating within a magnetic field to produce a voltage according to the relation

$$V(t) = NBA\omega\sin(\omega t)$$

where N is the number of turns of the rotating coil, B is the magnetic field, A is the cross-sectional area of the coil, ω is the angular frequency of the rotating coil, and t represents time. For a coil that rotates at constant speed in a uniform magnetic field, the voltage oscillates smoothly between maximum and minimum values according to a sine function. (See equation above.)

Electrical losses due to heating of transmission lines depend on the square of the average current in the lines. To minimize losses due to heating, utilities usually transport electrical energy at high voltages produced by high-voltage transformers. Turbine-generator systems generally supply electrical power for these high-voltage transformers. A transformer is a device capable of increasing or decreasing the voltage on its output, depending upon the number of turns on its secondary (output) coil versus its primary (input) coil.

A transformer increases (or decreases) the voltage on its output via magnetic induction, whereby a changing voltage applied to the input (primary) coil generates a changing magnetic field at the location of a secondary coil. A changing magnetic field present at a secondary coil produces a changing voltage within that coil. If the number of turns in the secondary coil is N_2 and the number of turns in the primary coil is N_1, the voltage on the secondary coil is (N_2/N_1) times the voltage on the primary coil. A step-up transformer has a ratio (N_2/N_1) that is greater than 1. A step-down transformer has a ratio (N_2/N_1) that is less than 1.

Transmission of electric power from generating stations to cities or towns occurs at voltages exceeding 300,000 V. Step-down transformers then reduce the electric potential for local distribution. Substations first reduce voltages to hundreds of Volts for transmission to primary users such as factories. Additional transformers further reduce voltages to 120–220 V for residential consumers.

TRANSMISSION LINES

Once electric power is ready for distribution, transmission lines carry that power over long conductors supported by central cables. Large towers placed at regular intervals allow the lines to run over large distances. The US grid now has over two million miles of power lines, along with 12,000 power plants and some 3,000 utilities.

Most of the installed transmission cables in the US use aluminum conductors with steel reinforcement. As temperatures increase within the conductors (with increasing current), the steel core expands, leading to more line sag between the support towers. Specifications limit temperatures of the conductors to 93°C for continuous operation (and higher temperatures for brief periods during emergency operations).

Advanced cable technologies employ composite cores made of carbon fiber and/or ceramics that offer added strength and less thermal expansion, as compared with steel. Thermal expansion coefficients for carbon fibers are one-tenth those of steel fibers, thereby reducing line sag. The added strength of carbon fibers allows smaller diameter cores, thus providing additional space for conducting material. Overall, advanced conductors convert less electrical energy into thermal energy, allowing generating stations to produce less electrical power to satisfy the demands.

DISCUSSION QUESTIONS

How Is the Grid Vulnerable?

As a vast and intricate machine, the grid is vulnerable due to problems, ranging from terrorist acts to any number of technical failings. Several extremist groups in the US have attempted or threatened sabotage of the electric grid. One of the most recent attacks occurred in 2022 in North Carolina where attackers breached gates at two substations and opened fire on electrical equipment, causing significant damage. Thousands of residents in the central part of the state did not have power for several days. This incident motivated the governor of North Carolina, Roy Moore, to announce a reward of $75,000 for information leading to the arrest and conviction of the person or persons responsible for the damage.

The grid is also susceptible to blackouts and brownouts due to large draws of electrical power from the system without appropriately compensating for these draws. A blackout occurs when the system has such high demands that components fail, and the grid voltage available to consumers falls to effectively zero. A brownout occurs when insufficient power is available on the grid to meet the demands, thus electrical grid voltages are lower than specified. Under these conditions, devices do not operate optimally or fail to operate altogether. To prevent such unwanted outages, utilities must quickly detect increases in demand to place electrical energy from other sources (or stored electrical energy) onto the grid. Other sources include backup generators or parts of the grid where excess electrical energy is available.

In addition to the unavailability of adequate power, transmission lines often fail due to fatigue or accidents. Occasionally, lines overheat due to high air temperatures or transmission of large currents for excessive periods of time. When lines overheat, they tend to sag under their own weight and ultimately contact a structure or possibly the ground.

What Are Some of the Key Parts of a Smart Grid?

The smart grid refers to various technologies that contribute to enhanced two-way communication between users and producers of electricity. Special meters track electrical usage on short time scales and sensors along transmission lines monitor demand and supply. Using these data, the grid system responds with the appropriate amount of electrical power. Other features of a smart grid include appliances that shift their usage to off-peak times and microgrids that supply reliable power to local customers. These smaller grids incorporate battery or other technologies to supply electrical energy when the broader system experiences outages (or other problems). Overall, a smart grid system increases efficiency and reliability, enhances responsiveness, and reduces outages.

What Are Some of the Most-Needed Grid Improvements?

To combat climate change, utilities need more electricity production from renewable energy sources such as wind, solar, and hydropower. Due to outdated infrastructure, however, transmission lines near proposed solar and wind sites often do not have sufficient capacity to carry the electric power produced by these installations. Such deficiencies cause cancellations or delays of renewable energy projects. Thus, one of the most needed grid improvements is installation of advanced conductors to handle the increased power from large wind and solar projects.

Another improvement requires installation of infrastructure to increase connectivity of all parts of the US grid system. Increased connectivity ensures that electrical power is accessible when disasters occur. For example, during the extreme cold snap of 2021 in Texas, unwinterized and not-properly-winterized generators failed, causing customers to lose power for many days. Residents resorted to stripping baseboards from their homes and burning them to stay warm. Transmitting excess electrical power from other regions proved impossible given the lack of connectivity between the main US grid and the Texas grid systems.

The grid system that serves most of Texas has few connections with the rest of the nationwide grid, thus failures of local and regional generating stations effectively isolate the state. One way to alleviate such problems in the future is to improve the transmission infrastructure so that the Texas system (called the Texas interconnection) and other vulnerable parts of the US grid have greater capacity for receiving electrical power. Implementation requires investment and willingness on the part of citizens and utilities in regions like Texas to increase their connectivity to the rest of the US grid system.

ADDITIONAL QUESTIONS FOR DISCUSSION

1. How will we finance or fund grid infrastructure improvements?
2. How does a modernized grid improve our daily lives?
3. What are the effects of an improved grid on the economy?

CHAPTER 6

The Nuclear Dilemma

T HE PUBLIC VIEW OF nuclear energy is one of apprehension, most likely due to horrific consequences of nuclear bombs and/or fear of hazardous nuclear wastes getting into the food supply or water systems. Indeed, nuclear bombs wreaked devastation on humanity and property when unleashed near the end of World War II. In addition, several nuclear accidents (in particular, Chernobyl in 1986) have caused thousands of human deaths, along with contamination that remains to this day. Discussions below acknowledge the dangers of nuclear fallout and spills, but, in addition, show how this strategic natural resource has the potential to provide humanity with hundreds of years of sustainable energy.

BACKGROUND

The idea that matter is made of basic and indestructible particles emerged from Greek thinkers, including Democritus. For centuries thereafter, workers viewed atoms as hard spheres packed together to form the materials that we see. Following the discovery of the electron in 1897, early models of atoms pictured a sea of positive charge with electrons embedded, much like seeds in a watermelon. Scientists referred to this version of the atomic model as the "plum pudding atom."

DOI: 10.1201/9781003511472-6

By 1911 Geiger and Marsden in collaboration with Rutherford showed that the "plum pudding model" could not explain certain results of various metal foil experiments. In these experiments, positively charged alpha particles bombard a thin metal (often gold) foil to produce deflections (scattering) of the incoming particles. To the researchers' astonishment, many of the alpha particles passed through the foil with little to no deflection (as if there is empty space within the atomic structure). In other cases, however, deflections occurred through large angles, including some in which the incoming particles were deflected backward. Astounded by this result, Rutherford wrote, "It was quite the most incredible event that has ever happened to me in my life. It was almost as incredible as if you fired a 15-inch shell at a piece of tissue paper and it came back and hit you."

With these results in mind, Rutherford proposed a new model in which positive charge occupies the core of the nucleus while negatively charged electrons orbit the nucleus. This planetary model explains the basic results of the metal foil experiments: alpha particles shot directly at nuclei reverse direction upon close encounter, ones passing near nuclei undergo large-angle deflections, and those passing through the volume between nuclei are largely unaffected.

While the planetary model explains the basic scattering of charged particles bombarding atoms, it does not explain two other important atomic effects. The first is that atoms emit only certain frequencies of waves but not others. (Frequency for waves refers to the number of oscillations that occur in a fixed amount of time.) The second is that electrons do not emit energy as they orbit around the nucleus as expected from classical physics. In classical models, charges undergoing acceleration during their orbits emit energy and eventually spiral into the nucleus. Both effects require quantum-mechanical models of atoms.

NUCLEAR PHYSICS

With the discovery in 1896 of mysterious emissions from uranium compounds, scientists sought to understand the nature of these unknown rays (radiation). Interactions with magnetic fields showed that the rays were of three types: alpha, beta, and gamma. Subsequent experiments verified that alpha rays are positively charged helium nuclei, beta rays are electrons, and gamma rays are high-energy photons.

Following the scattering work described above and evidence that electrons occupy orbits around a positively charged core, scientists proposed an early nuclear model. In this model, the central core (called the nucleus) contains protons and neutrons, commonly known as nucleons. In turn, these particles consist of fundamental particles referred to as quarks. Neutrons have one up quark and two down quarks; protons have two up quarks and one down quark. Quarks determine several properties of sub-atomic particles, including mass and charge.

Nuclear models also predict nuclei that are dense and small with a radius of order 10^{-15} m. (For perspective, a trillion nuclei side-by-side would fit on the head of a pin.) Close packing of protons and neutrons into the atomic core is somewhat unexpected, because protons repel one another due to their like positive charges. These Coulombic forces also cause attraction between two objects rubbed together such as your hair and a comb given that they obtain opposite charges.

At close range, however, all nucleons experience extremely large attractive forces due to the strong nuclear force. The strong nuclear force is one of four known fundamental forces in nature and has a relative strength of 10^{38} as compared to gravity. (This means that two nucleons inside a nucleus would experience a force that is 10^{38} times stronger than the gravitational force between the same nuclei.)

In nuclei where the strong force dominates, nuclear lifetimes are extraordinarily long, and the nuclei are stable. Light nuclei are most stable when the number of neutrons is equal to the number of protons. Heavy nuclei, by contrast, are most stable when the number of neutrons exceeds the number of protons. Heavier nuclei require extra neutrons to produce sufficient attractive forces to keep the nuclei together. Otherwise, the large number of proton–proton repulsive interactions cause the nuclei to separate.

Ultimately, the large number of repulsive forces between pairs of protons cannot balance the extra attractive forces between neutrons, so nuclei with atomic number Z (number of protons) greater than 82 become unstable. For example, all nuclei (isotopes) associated with uranium ($Z = 92$) are unstable.

Unstable nuclei spontaneously decay by one of three processes: alpha decay, beta decay, and gamma decay. In the process of alpha decay, the original nucleus (known as the parent) loses two protons and two neutrons to produce a daughter nucleus plus a helium nucleus, containing the four lost particles. Generally, daughter nuclei are more stable than parent nuclei; however, further decays often produce the most stable and long-lived nuclei. Once stable nuclei form, the chance of additional decays is small.

MODELS OF NUCLEI

Following discoveries of the structure of the atom, scientists focused on developing more extensive models of the nucleus. One of the goals was to explain nuclear binding energy. When assembled, the overall mass of the nucleons in a nucleus is less than the total mass of the individual nucleons. This decrease in mass means the assembled nucleus has less overall energy due to mass-energy conversions. (The expression $E = mc^2$ describes this transformation.) As a result, separating the nucleons within a nucleus into individual particles requires addition of energy equivalent to this binding energy.

First proposed by Bohr in 1936, the liquid drop model assumes that nucleons behave as if they are molecules in a drop of liquid. Like molecules in liquids, nucleons within nuclei interact strongly with each other and experience frequent collisions. The main interaction between nucleons in the nucleus is a strong attractive force (akin to cohesive forces in liquids). Data indicate that the binding energy per nucleon is nearly constant, thus the binding energy itself is proportional to the number of nucleons, referred to as the mass number A. Given that the volume of the nucleus is proportional to A, the liquid drop model has a term in its binding energy determined by A. This dependence is known as the volume effect.

A second interaction within the liquid-drop model contributes to the binding energy due to surface effects. Some of the nucleons occupy locations (sites) near the surface, thus producing weaker forces binding these nuclei to the drop (due to fewer neighbors). The overall effect is that the surface nucleons reduce the overall binding energy by an amount proportional to the surface area of the drop.

Repulsive forces between protons in the nucleus due to Coulombic forces also reduce the binding energy of the nucleus. The potential energy associated with the Coulomb force is proportional to the number of pairs of nucleons and is inversely proportional to the nuclear radius. Thus, an increase in the number of pairs of nucleons causes the nuclear binding energy to decrease and the nucleus to become more unstable. This instability leads to nuclear decay as described above. By considering the three interactions discussed here (and an additional term for heavy nuclei with many excess neutrons), the liquid-drop model accurately predicts binding energies.

Another model used to describe the nucleus is the independent-particle or shell model. Here the assumption is that nucleons act as particles that occupy orbitals and shells, much like electrons in atomic models. Orbits have well-defined energy levels, often pictured as rungs on a ladder. Proton energy levels are slightly higher than corresponding neutron energy levels due to proton–proton interactions causing repulsion. Protons and neutrons are spin 1/2 particles; thus, they possess one of two spin states. (Here, think heads or tails for coins.) Quantum mechanics rules specify that nucleons with the same spin states cannot occupy the same energy level. As a result, each energy level has one "up" state and one "down" state.

The probability of an "up" spin is the same as a "down" spin, so the energy levels fill from the ground level, essentially two by two, until all the particles occupy specific energy levels. Addition of a lone proton or neutron requires that particle to occupy a higher energy level, thereby increasing the overall energy of the nucleus. Given this added energy, those nuclei that have an even number of protons and neutrons are more stable than others. Approximately 160 stable isotopes with an even number of protons and an even number of neutrons (called even–even nuclei) are known, while only four stable isotopes with an odd number of both nucleons are known. The shell model is particularly useful for explaining excited states in nuclei and the presence of quantized states (nuclear energy states with specified values).

NUCLEAR FISSION

The nucleus is a vast repository of stored energy, released via nuclear reactions. One process by which this occurs is nuclear fission. In fission reactions, a nucleus of large mass splits to form smaller nuclei and releases an amount of energy determined by the loss of mass during the reaction. Otto Hahn and Fritz Strassman first observed nuclear fission of uranium nuclei (^{235}U) in 1939, following basic studies by Fermi.

The breakup of the uranium nucleus occurs due to a sequence of events. To initiate the reaction, the ^{235}U nucleus first captures a slow-moving neutron to form an excited ^{236}U* nucleus. This new nucleus has excess energy that causes nuclear oscillations, thus distorting the shape of the nucleus. Once distorted the nucleus looks like a dumbbell. In this scenario, enlargement of the volume allows proton–proton repulsive interactions to dominate and ultimately produce splitting of the nucleus. (Recall that in the liquid-drop model the nucleus behaves as a drop that acquires excess energy, becomes distorted, and separates due to large amplitude oscillations.)

The fission process (namely of ^{235}U here) yields two fission fragments, several neutrons, and energy. Each fission event produces 200 MeV of energy—equivalent to an amount needed to increase the speed of a dust particle from 0 to 300 m/s. Fission processes account for approximately 10% of the world's annual electricity production. Most nuclear bombs employ fission reactions that release large quantities of energy over short time scales.

NUCLEAR FISSION REACTORS

The release of neutrons during nuclear fission allows for the possibility of self-sustained reactions—ones that continue without the need for additional nuclear fuel. In such cases, uranium nuclei capture released neutrons from other fission events, leading to splitting of additional uranium nuclei. When controlled, these sustained chain reactions continuously produce energy at a desired rate.

A nuclear reactor is a system, designed to maintain self-sustained nuclear reactions. Nuclear reactors generally operate using enriched uranium as fuel. (Enrichment refers to the process of increasing the concentration of nuclei needed for fission.) Enrichment is necessary because natural uranium only contains about 0.7% ^{235}U, the isotope that undergoes fission reactions. The other main isotope of uranium (^{238}U) does not undergo fission but absorbs neutrons, leading to the production of unwanted neptunium and plutonium. Fission reactions of ^{235}U generate an average of 2.5 neutrons during each fission event. To achieve sustained reactions, a fissionable nucleus must absorb one of these released neutrons (on average) and then undergo fission.

Designers and engineers characterize reactors by a value known as the reproduction constant K. The reproduction constant is the average number of neutrons from each fission event that cause another (fission) event. When $K = 1$, the chain reaction sustains itself, and the reactor is classified as critical. When K is less than 1, the reactor is subcritical, and the reaction does not continue over time. When K is greater than 1, the reactor is supercritical, and the reaction is uncontrolled. To produce energy for a utility, a nuclear reactor must operate with a K near unity.

The central workhorse of a nuclear reactor is the core, the region where the nuclear fuel resides, and the reactions occur. To maintain those reactions and provide maximum safety, the reactor core consists of four main parts: the nuclear fuel elements, control rods, moderator material, and radiation shield. The fuel elements contain enriched nuclear fuel where the reactions occur. Control rods provide the main control of K by absorbing enough neutrons to keep the reactor self-sustaining.

Nuclear operators partially or completely remove control rods to increase the average number of available neutrons when K becomes too small. By contrast, workers insert control rods farther into the reactor or insert additional ones when K becomes too large. In addition to controlling the number of captured neutrons, the speed of those neutrons must fall within a certain window to ensure capture of neutrons by fissionable nuclei. Materials such as water are ideal for slowing neutrons to initiate capture. Beyond the inner core a containment (radiation) shield absorbs neutrons that escape the core to prevent damage to other parts of the reactor and protect human workers.

One of the nuclear reactor designs, used to produce electrical energy, is the pressurized-water reactor. Here the reactor core works in concert with a closed water system, turbine, and generator. Energy released during fission events in the reactor core heats water within a primary loop, which is in contact with water in a secondary loop via a heat exchanger. Water in the secondary loop converts to steam, which drives a turbine/generator to produce electricity. Used steam from the turbine enclosure then enters a condenser where the steam encounters a cold loop. After condensation of the steam, the liquid water enters the heat exchanger again to produce more steam.

Reactor safety is an obvious concern for all. Interruption of water flow around the reactor core due to the failure of pumps or leaks in the water system is of highest concern. In the worst cases, rising temperatures cause melting of the fuel elements and ultimately melting of the bottom of the core and ground below. Such a disaster would not only destroy the reactor but also release radioactive materials into the ground and air. To prevent these kinds of tragedies, nuclear reactors have backup cooling systems that are available if the main ones fail.

Disposal of radioactive waste is another concern when operating nuclear reactors. The spent fuel contains long-lived, highly radioactive isotopes, requiring safe storage over time to allow the activity (decays per second) to decrease to safe levels. One option is to transport the nuclear byproducts to remote locations such as abandoned mines or mountain enclosures for storage. To date this solution is not viable given the public outcry regarding transport of nuclear waste. The public is wary of shipping nuclear waste on rail and roadways where accidents potentially cause leaks of toxic materials. The irony is that many folks consume electrical energy produced by reactors in their areas of the country, but they oppose transport of wastes from those reactors through their cities or towns.

Sabotage of nuclear power plants is the worst nightmare of public officials and those working in the nuclear industry. No single scenario is the most likely one, so extreme vigilance is essential. Those who desire harm would most likely attempt to breach the reactor core to release radioactive materials—not only causing harm due to the presence of toxic materials but also fear due to the thought of nuclear materials in the environment. Operation of nuclear reactors also requires careful management of thermal energy, produced as water and other materials interact with the reactor core. Condensation of steam, originally used to turn the turbine, releases excess energy via large cooling towers, thus adding significant infrastructure to reactor designs.

NUCLEAR FUSION

Nuclear reactions in which two lighter nuclei join to form a heavier one are known as fusion. The interior of stars (like our Sun) has ideal temperatures and pressures to fuse hydrogen into helium. Formation of stars occurs in regions of space that have vast clouds of dust and gas. These materials likely originate from previous generations of stars that have exploded or leaked material during their demise as main sequence stars. In certain regions of the gas and dust clouds, shock waves compress the particles to the point that they collapse under the action of gravity. As collapse occurs, the particles move with greater and greater speeds thus raising temperatures within those regions.

Once temperatures are in the range of 10^7 K and densities are sufficient to maintain certain collision rates, self-sustained fusion reactions occur, and the star maintains equilibrium. In equilibrium, the inward force of gravity and outward forces due to thermal effects (such as convection) and radiation pressure balance one another. Radiation pressure occurs due to momentum carried by light waves emanating from the star's core. Once at equilibrium, the star maintains fusion reactions for billions of years.

As the star produces energy via fusion reactions, hydrogen converts into helium with the concomitant production of gamma rays, positrons, and neutrinos. Neutrinos rarely interact with matter, so they carry energy from the star across vast reaches of the universe. As gamma rays move outward from the core and interact with stellar material, they convert to lower-frequency photons in the ultraviolet, visible, infrared, and other regions of the electromagnetic spectrum.

NUCLEAR FUSION REACTORS

The study of stars reveals many insights into the physics of nuclear fusion. Moreover, discoveries in nuclear engineering suggest that nuclear fusion has the potential to meet the energy needs of future generations. No practical, continuously working fusion reactor is yet available, despite several advances over the past few years. Conceptually, a fusion reactor mimics the Sun to fuse hydrogen into helium, thus releasing significant amounts of energy per each fusion reaction.

Fusion reactors represent the ultimate energy source given that their basic fuel is water. All proposed fusion reactors use deuterium as fuel rather than the lighter hydrogen isotope (^1H), because the lighter isotope requires pressures and densities comparable to those inside the Sun to achieve fusion. Extraction of heavier nuclei from ordinary water produces the required deuterium. The byproducts of fusion are nonradioactive and inert.

The main impediment for Earth-based fusion is overcoming Coulombic repulsive forces between (positively) charged nuclei. One way to achieve conditions needed to induce fusion is to raise temperatures within the reactor core. Such extreme temperatures are difficult to produce; moreover, they generate a soup of ionized particles known as a plasma. Fusion reactions require sufficient plasma ion density and plasma confinement time (in addition to high energies) to ensure production of more fusion energy than necessary to heat the plasma.

The most promising confinement method for nuclear fusion consists of two magnetic fields that keep the nuclear plasma within a region shaped like a doughnut. Present fusion reactors require additional energy to initiate fusion. Injection of neutral energetic particles into the plasma provides this extra energy. Once nuclear fusion energy becomes practical, it offers several advantages over nuclear fission: abundant and relatively cheap fuel requirements (deuterium produced from water), the absence of weapons-grade material as byproducts, and reduced radiation hazard.

RADIOACTIVITY AND HUMANS

The release of energy and/or particles by alpha, beta, and gamma decays is a threat to humans and other living organisms, primarily due to radiation interacting with bodily cells. In the most extreme cases, the organism dies in a few hours due to large doses destroying many cells. Lesser levels of radiation initiate mutations of cells that accumulate over time, ultimately causing cancer or other debilitating diseases.

Radiation damage in matter depends on the properties of the absorbing material, along with the type and energy of the radiation. In biological systems, radiation damages cells mainly due to ionizing effects in which electrons separate from molecules. Upon removal of electrons, highly reactive ions or radicals often form. For example, interactions between radiation and water molecules produce hydrogen and hydroxyl radicals. These kinds of radicals induce chemical reactions that break bonds in proteins and other vital molecules.

In addition to the production of ions and radicals, ionization effects impact cells. If enough molecules within a cell become ionized, cell death occurs, resulting in permanent damage to the organism or even death. Cells that survive ionization, often experience damage (mutations) that have cumulative effects (as mentioned above). Effects involve both somatic and genetic damage. Somatic damage refers to harmful effects to nonreproductive cells in biological systems, whereas genetic damage refers to harmful effects to reproductive cells.

Quantitative effects of radiation on the human body require the amount of radiation received, along with the effectiveness of that radiation. Scientists generally use units of rad (radiation absorbed dose) to express radiation dose. One rad refers to the amount of radiation that deposits 10^{-2} J of energy into 1 kg of absorbing material. The effectiveness of radiation in causing biological damage varies according to the type of radiation (and some other factors). The RBE (relative biological effectiveness) factor refers to the number of rads of x-radiation or gamma radiation that produces the same biological damage as 1 rad of the radiation in use. As an example, a given dose of alpha particles produces 10–20 times more biological damage than an equal dose of X-rays. These two factors combine mathematically to obtain the dose in rem, as shown below (see also Table 6.1):

$$dose\ in\ rem = dose\ in\ rad \times RBE.$$

A one-time dose of less than 100 rem causes no immediate bodily symptoms. Cells largely repair themselves by natural processes, or the

TABLE 6.1   Type of radiation and the RBE factor of each

Type of radiation	RBE factor
X-rays and gamma rays	1.0
Beta particles	1.0–1.7
Alpha particles	10–20
Slow neutrons	4–5
Fast neutrons and protons	10
Heavy ions	20

body adjusts to damage without affecting normal functions. (Over time low doses statistically increase the risk of cancer as we will discuss later.) A 200-rem dose causes radiation illness in which the body experiences nausea, listlessness, and fatigue. These symptoms are essentially the ones experienced by patients undergoing cancer treatments.

Higher doses of radiation cause irreparable damage to the body. At 300 rem the chance of death is 50% (referred to as LD50, lethal dose 50%). Exposure to 500 rem all at once without medical treatment likely causes death due to effects of the radiation. Exposure to 1,000 rem causes incapacitation within hours and certain death.

Another way to express doses in rem is to specify a certain number of decays of a particular kind of radiation. For example, approximately 10 trillion gamma rays produce 1 rem. For perspective, persons less than a mile away from the epicenter of the 1945 Hiroshima bombing experienced absorption of 9.46 Gy. This level of radioactivity equates to 946 rem and is double the amount needed to cause death due to whole-body exposure.

Beyond the effects of one-time exposure to radioactivity, continued exposure increases the risks of cancer and other diseases due to cumulative effects. From data collected on cancer rates, the increased risk of cancer due to ionizing radiation follows a linear dependence when plotted against dose in rem. Simply said, the risk of cancer from radiation increases in proportion to the amount of rem received by the body. (This linear effect has an upper limit of 2,500 rem, considered to be a cancer dose.) Figure 6.1 shows this effect.

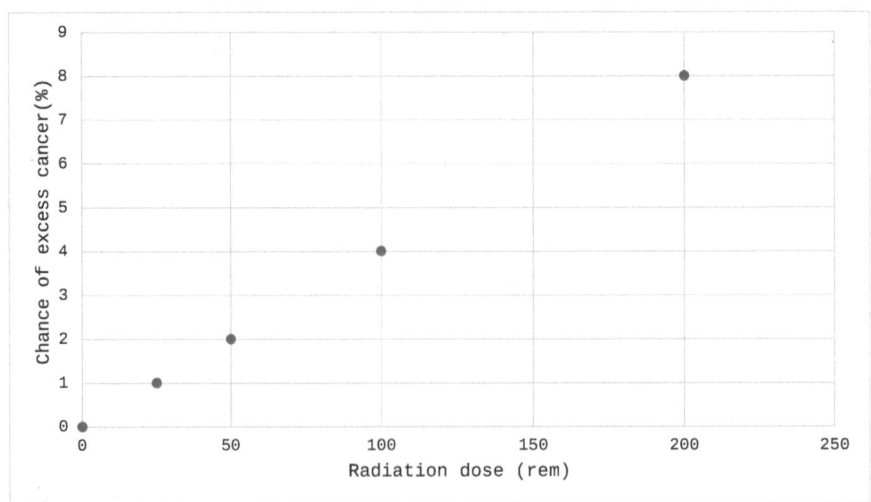

FIGURE 6.1   Increased risk of cancer versus radiation dose

NUCLEAR WEAPONS

A uranium bomb concentrates enough fissionable material to obtain a supercritical mass. To build such a bomb, acquisition of enriched uranium is essential. Purchase of enriched nuclear material violates international law, but rogue nations and groups typically show no regard for laws or norms. Enrichment efforts—either lawful or otherwise—produce nuclear fuel using methods such as gas diffusion and centrifugation. The International Atomic Energy Agency (IAEA), sanctioned by the United Nations, monitors nuclear enrichment worldwide. The goal of the IAEA is to promote peaceful use of nuclear energy. Monitoring of countries like Iran is essential to prevent unlawful actions and to avoid military escalations.

One of the simplest nuclear weapons consists of two subcritical masses of ^{235}U separated within a nuclear device. This design prevents a runaway chain reaction given the separation between the two compartments. Detonation of the device forces the two subcritical masses together to generate a supercritical mass that results in a massive explosion along with fallout of radioactive material from the bomb.

A plutonium bomb also generates a runaway nuclear reaction using ^{239}Pu as the nuclear fuel instead of ^{235}U. Plutonium is available as a byproduct of nuclear reactors; however, plutonium reactions are prone to pre-detonation due to the presence of ^{240}Pu nuclei. This isotope (heavier nucleus) fissions spontaneously without the need to capture a neutron. Release of these neutrons before detonation causes partial chain reactions to occur, thus causing the bomb to explode prematurely without releasing much energy. To avoid difficulties with pre-detonation, early bomb makers devised an implosion technique when plutonium is used as the nuclear fuel.

An implosion bomb uses a shell of dispersed plutonium surrounded by explosives. When activated the explosives drive the plutonium fuel inward (toward the center), causing significant compression. This extreme compression of the plutonium fuel produces a nuclear chain reaction without premature explosion.

DISCUSSION QUESTIONS

Is Leakage of Radioactive Materials from Nuclear Reactors a Major Problem?

Under normal operation, nuclear reactors are entirely safe and have multiple redundancies to prevent disasters. For example, if the main cooling water pump for the reactor fails, a backup pump is available.

However, like much of our infrastructure, nuclear reactors are susceptible to terrorist attacks. If an attack occurs, the main threat is damage to the reactor and/or the cooling systems. Significant breaches of the reactor obviously cause leakage of nuclear material and possible harm to humans and the environment. Loss of the cooling systems in worst-case scenarios results in a reactor meltdown (melting of the reactor, its radiation shield, and possibly the bottom of the reactor). Meltdowns often result in release of radioactive material due to damage to parts of the reactor that house the nuclear fuel. Destruction of the turbine or generator of the nuclear power plant would be no different than if an attack occurred on comparable parts of a coal-fired or natural gas power plant.

As observed during the 9/11 attacks, terrorists show no regard for their own or others' lives and often resort to heinous acts to create chaos, confusion, and destruction. To prevent these tragedies, vigilance on the part of public officials, workers, and citizens is crucial. As discovered during investigations of the 9/11 attacks, one of our greatest failings in thwarting the horrific events of 9/11 was a "failure of imagination." We simply failed to imagine the methods of destruction devised by those seeking to inflict harm.

Should the United States Establish a Policy of Storing Nuclear Waste in Offsite Locations Away from Nuclear Power Plants?

The storage of nuclear waste in offsite locations is the most prudent approach for two reasons. Onsite storage of nuclear waste is not as safe as in more secluded locations such as the one proposed at Yucca Mountain. Storage in remote locations inside well-protected formations prevents serious harm to humans from leaks or breaches. On the contrary, most onsite storage facilities are inside buildings that are vulnerable to terrorist attacks and are closer to humans. In addition, storage of waste in remote locations offers the possibility of managing the waste more effectively. For example, when nuclear materials become less active (in terms of their radioactivity) they can be moved to less protected areas to make room for the more active materials.

One of the obvious reasons for not storing materials offsite is the problem of transport. Many folks fear the possibility of nuclear spills during transport by rail or other means. At present, no plans to store nuclear waste in offsite locations are under consideration in the United States.

Is a Dirty Bomb Dangerous?

A dirty bomb is a conventional bomb with radioactive material packed within the bomb casing. When exploded, nuclear material disperses over a region defined by the debris field of the bomb. Small debris fields generally affect fewer people but produce higher dosages due to increased concentrations of nuclear materials. Harmful dosages are unlikely, but close monitoring is necessary to ensure safety. Dispersed nuclear material poses no significant physical threat. All dirty bomb explosions require experts to clean up the material and monitor the affected areas. In most cases, harm is minimal, especially if no one dies during the initial blast.

The most significant effect of a dirty bomb is its psychological impact. Humans have a natural fear of what they cannot see, so the thought of the presence of invisible "rays" or radiation weighs on our minds. Following such an attack, public officials must assure citizens that long-term effects are minimal. A comprehensive plan requires proactive efforts to prevent such attacks, ample resources for clean-up and monitoring, and effective campaigns to inspire public confidence.

Is It Wise for Cancer Patients to Take Radiation Therapy?

For many folks the thought of exposing their bodies to radiation is extremely scary. These feelings are understandable given that nuclear radiation in low doses causes sickness and, in larger doses, certain death. Despite these hazards, medical experts view radiation as a viable way to destroy cancer cells and thus rid the body of cancer. Cancer cells multiply very quickly but do not have the means to repair themselves. Thus, radiation is effective in killing cancer cells.

By contrast, normal cells largely repair themselves when exposed to radiation in relatively low doses. Some of the newest treatment techniques deliver energy to targeted areas, thus ensuring maximum effects. Evolving radiation delivery or other techniques selectively destroy cancer cells while sparing normal cells. Overall, radiation therapy continues to save lives and deserves strong consideration when assessing cancer treatment.

What Are the Advantages and Disadvantages of Adding New Nuclear Power Plants?

Nuclear power plants produce electrical energy that is reliable, clean (from the standpoint of CO_2 emissions), and sustainable over many hundreds of years. Assuming current usage, world-wide uranium reserves have projected lifetimes of at least 200 years, assuming undiscovered reserves have capacities close to expectations. Nuclear power plants also generate significant amounts of energy when considering the size they occupy. Typical nuclear power plants such as the one in Callaway County, Missouri, produce over a GW ofelectrical power.

Obvious disadvantages of nuclear energy include production of dangerous wastes, the need to continue to mine uranium ore and/or purchase it from other countries, and public perception that nuclear energy is inherently dangerous. Despite these disadvantages nuclear energy offers a clean alternative to fossil fuels. France generates around 60-70% of its electricity needs from nuclear power plants. Solving the problem of nuclear waste storage in the United States is paramount for increasing our nuclear power capabilities.

CHAPTER 7

Our Place in the Universe

MODERN STUDIES OF OUR universe reveal that the Earth is not centrally located as the ancients believed nor is it unique in composition or form. Rather, the Earth is a rocky planet with seven local neighbors, all of which orbit an ordinary star near the edge of a vast collection of other stars comprising the Milky Way galaxy. Billions of other galaxies of various types occupy the observable universe. From all available evidence, the universe itself is expanding, and that expansion is accelerating.

While the Earth is ordinary as observed from afar, it boasts millions of species and has oceanic water covering 71% of its surface. The presence of vast amounts of liquid water and a stable atmosphere support numerous life forms, including those that can shape their environment and make ethical decisions. Other locations within our galaxy have planets that reside close enough to their central stars to possess liquid water, but, thus far, scientists observe no evidence of advanced life forms. Efforts to determine if life exists beyond Earth continue.

The universe originated from a single event known as the Big Bang. From this initial state of high density and temperature emerged spacetime along with sufficient energy and matter to form the known universe. Early in the evolution of the universe, various atoms developed from elementary particles. Hydrogen and helium materialized as the first elements;

DOI: 10.1201/9781003511472-7

thus, upon formation, first-generation stars only contained these two. As some of these stars completed their life cycles and experienced stellar death, they added heavier elements, including carbon, oxygen, silicon, and iron, to those available for generating new astronomical bodies. Today, new stars form within regions known as the interstellar medium, mid-life ones produce constant energy as they undergo nuclear fusion during their long lifetimes, and older stars spew materials into this medium as they end their lives.

INTERSTELLAR SPACE

While the universe has an abundance of stars, planets, and larger structures such as galaxies, it also has regions between the stars and planets with densities a trillion, trillion times less than those associated with stellar or planetary bodies. These vast regions, known as interstellar space, contain as much matter as the stars themselves. Here new stars form, and dying ones eject debris that serve as the raw materials for new astronomical bodies (formations).

Interstellar matter is made of dust and gas, composed primarily of individual atoms with sizes of order 10^{-10} m or less, along with some molecules no larger than 10^{-9} m across. Many parts of the electromagnetic (EM) spectrum, including ultraviolet, visible, infrared, and radio waves, easily pass through interstellar matter due to the small sizes associated with the gas particles.

Interstellar dust exhibits more robust behavior. The densest accumulations of dust particles effectively block light due to scattering. Scattering of light generally depends on the size of the particles ("scatterers") and the wavelength of light. Terrestrial experiments confirm that dust particles in the interstellar medium are approximately 10^{-7} m, which is comparable to the wavelengths of many parts of the EM spectrum. (This dimming of light passing through the medium is referred to as extinction.) Overall, short-wavelength radiation is opaquer than long wavelengths, thus light from distant stars appears redder than is the case. Radio waves are long enough to penetrate and pass through the interstellar medium, thus making them effective in studying vast reaches of the universe.

Direct observations of the interstellar medium require techniques that detect weak radiation from atoms occupying that space. Much of this work focuses on radio emissions from cold, dark atoms, particularly hydrogen. Atomic hydrogen consists of a central proton with one electron occupying a spherical cloud around the nucleus. As elementary particles, protons and electrons possess a property known as spin. (Loosely speaking, spin is associated with particles spinning about their axes.) Both particles can exist in one of two spin states known as up and down.

When both particles (proton and electron) have the same spin states, the atom has a slightly higher energy than when the particles are in opposite spin states. As one of the particles changes to the opposite state, the atom emits a photon with specified energy and transitions to the lower energy state. Radiation from cold hydrogen atoms occurs mainly due to this process; this radiation corresponds to a frequency of 1.4 GHz (or a wavelength of 21 cm).

Energy differences between the two spin configurations are comparable to ones associated with temperatures of order 100 K. Thus, collisions between hydrogen atoms in the interstellar medium continually promote a fraction of hydrogen atoms to the higher spin states. These conditions guarantee that all parts of the interstellar medium emit 21 cm radiation. Frequency shifts of these emission lines reveal how fast clouds of atoms move relative to an observer. Knowing directions of motions and their magnitudes provides a wealth of data for monitoring the state and evolution of interstellar clouds.

Temperatures within the interstellar medium depend on their proximity to stars or other sources of radiation. Regions far from sources of radiation have temperatures of a few Kelvins, while regions near sources have temperatures of a few hundred Kelvins. Along with low temperatures, interstellar regions have extremely low densities, averaging approximately 10^6 atoms per cubic meter.

THE SOLAR SYSTEM, STARS, AND BIOCHEMICAL MOLECULES

The precise conditions associated with the early solar system are unknown; however, likely scenarios suggest that planetary systems form from a large interstellar cloud, composed of hydrogen and helium atoms, along with some heavy elements contained within gas and dust. Contraction of such a large cloud originates with the passage of another interstellar cloud or explosion of a nearby star. Once begun, the cloud contracts, rotates more quickly, and becomes a flat disk when its size reaches about 100 AU. (1 AU is the average Earth–Sun distance.) From observations, such disks exist around other stars, thus providing evidence for the theory of formation from a rotating interstellar cloud.

Formation of planets begins within the rotating disk by a process of accretion whereby particles collide and stick to one another to ultimately generate objects measuring a few hundred kilometers across. Once these infant planets become this size, they accumulate additional matter at an increasing rate due to effects of gravity (without the need for chance collisions). During this stage, the rate of growth of the masses increases further. In addition, this nascent solar system now has millions of objects the size of small moons, referred to as planetesimals, along with substantial amounts of hydrogen and helium.

As accretion continues and enters a second phase, planetesimals collide due to gravitational forces and stick together to form larger and larger objects. With this growth planetesimals increase their gravitational effects, thus causing attraction of other planetesimals. As a result, planetesimals combine to form protoplanets with little matter between them. Orbits then become more well-defined, circular in shape, and more widely spaced. In some models, the larger planets accumulate additional leftover matter, consisting of hydrogen and helium while the smaller planets do not accumulate gaseous matter.

Within our solar system, Jupiter, Saturn, Uranus, and Neptune, known as gas giants, consist of outer layers of hydrogen and helium. The inner planets—Mercury, Venus, Earth, and Mars—known as terrestrial planets, do not have outer layers of gas. As protoplanets grow, another process known as fragmentation occurs. Here, strong gravitational forces initiate high-speed collisions that break larger masses into smaller ones. Protoplanets then capture these smaller pieces, or the particles separate to become part of the system of comets and meteors known to contribute to the development of the infant solar system.

Temperatures partially determine the location of Jovian versus terrestrial planets within the rotating disk. As the large disk contracts, temperatures near the center increase more so than the outer regions. In the hotter, central regions, dust particles break into smaller and smaller particles, eventually forming ionized atoms. Thus, in the central core, where the protosun forms, hydrogen and helium are prevalent. Indeed, hydrogen and helium are the major components of stars like the Sun when they first form.

Regions of the disk beyond the protosun cool as gases radiate their energy. In these cooler regions, dust particles reemerge by condensation (much like water droplets condensing from water vapor on cold mornings).

The types of materials that condense depend on temperatures in that region of the developing solar system. In regions closer to the protosun, only metals condense. As temperatures decrease, additional molecules like silicates form at distances of 1 AU (the average Sun–Earth distance) from the protosun, followed by water ice at 3–4 AU and ammonia ice at 7–8 AU. Thus, planets like Earth have high concentrations of silicates and metals. By contrast, gas giants form from condensed ices (water and ammonia), along with condensed metals and silicates, primarily in their cores. Models suggest that the giant planets grow more quickly due to the availability of various kinds of condensates.

While the condensation model explains many features of the solar system, several observed anomalies require explanations beyond the current one. For example, Uranus has a rotation axis that is tilted quite substantially relative to the other planets. Here scientists theorize that a grazing collision caused Uranus' orbit to tilt. Such theories fall into the category of catastrophes. Another dilemma centers around the so-called angular momentum problem in which the Sun has very little of the solar system's angular momentum but nearly all its mass. Theories suggest that the solar wind or ejected planetesimals carried away the Sun's large angular momentum.

The condensation model is consistent with formation of a central star within the disk of an interstellar cloud that undergoes rotation and contraction. As the central region collapses due to gravitational forces, temperatures near the core of the cloud reach sufficient levels to initiate nuclear burning (nuclear fusion), and contraction ceases until later in the star's life cycle. While the birth of a star seems inevitable, several factors compete with gravity to prevent formation. At temperatures greater than absolute zero all particles undergo motions that prevent clusters of atoms from remaining together after collision. Clusters that form temporarily often disperse quickly due to motional effects.

Computer models suggest that approximately 10^{57} atoms at cloud temperatures of 100 K are necessary to cause gravitational collapse under ideal conditions. Complicating star formation via gravitational collapse are effects due to rotation of the disk and the presence of magnetic fields. When a spherical cloud rotates, it tends to bulge along its equator. As contraction continues, the disk rotates more quickly thus leaving weakly held particles behind. Magnetic fields also compete with disk contraction given that charged particles within the disk tend to spiral around field lines. These interactionsrestrict motions of charged particles, thus preventing sufficient contraction needed for stellar and planetary evolution.

Galaxies probably developed soon after the Big Bang as clumps of matter formed under the action of gravity. Some of these clumps broke into pieces, but others accumulated more matter that eventually collapsed to form an infant galaxy. Further evolution produced what we observe today. The cycle of birth, death, and rebirth continues within each galaxy in the universe.

In 1953, Harold Urey and Stanley Miller performed the first experiments to determine if complex molecules develop from simpler ones under certain conditions. In their closed-loop arrangement, water boils in a lower flask, causing water vapor to migrate to a vessel consisting of water, methane, carbon dioxide, and ammonia gases. An electrical discharge within the discharge tube energizes the gases along with the water vapor from the flask. Contents from the discharge tube then condense and enter a trap from which contents accumulate for analysis later. The apparatus runs continuously in this closed loop for about 1 week. Results of these early experiments show formation of amino acids during this process.

Later experiments proved that amino acids joined at elevated temperatures produce protein-like blobs. These entities cluster to form microspheres that allow passage of small molecules, which react to form more complex molecules. Microspheres then prohibit passage of these more robust molecules, causing the spheres to grow. Over time the spheres split to form smaller droplets.

The development of life as we know it certainly required many stages of evolution in which higher and higher forms appeared; moreover, the Urey experiments (and subsequent work) proved that the universe is potent enough to produce the building blocks of life. Today biologists contend that living things emerged via an evolutionary process from a common ancestor that lived about 3.5–3.8 billion years ago. Natural selection produced successive generations, adapted to the conditions for those specific eras. Over time the repertoire of life forms increased in diversity, yielding the myriad of organisms observed today.

ARE THERE LIFE FORMS COMPARABLE TO HUMANS IN OTHER PARTS OF THE UNIVERSE?

No clear evidence exists to confirm life forms comparable to humans anywhere in the universe, although much speculation concerning extraterrestrial life continues, both in the mainstream and alternative media. The search for Earth-like planets gained traction in 1992 with the confirmation of several terrestrial-size exoplanets orbiting the pulsar PSR B1257+12. Additional searches culminating in 1995 revealed a giant planet orbiting 51 Pegasi. In 2023, an Earth-like exoplanet LHS 475 b was discovered by the James Webb Space Telescope. To date more than 5,000 exoplanets are known.

To support life as we know it, a planet must develop a stable atmosphere, maintain temperatures above the freezing point of water, and provide protection from quickly moving charged particles and other incoming radiation. Planets within such regions could conceivably sustain carbon-based life forms like those on Earth. While such conditions are ripe, they do not guarantee that advanced life evolved in other places within the universe. Evolution of advanced life requires not only the right conditions but exquisite timing to ensure that organisms survive and propagate.

Another way to determine if advanced life exists is to monitor radio and other EM waves to see if any discernible communication patterns emerge. The general collection of efforts that look for evidence of intelligent civilizations sending signals into space is known as SETI (Search for Extraterrestrial Intelligence). One active effort known as METI (Messaging to Extraterrestrial Intelligence) involves sending signals into space with the goal of intelligent life detecting them and possibly replying.

TO WHAT EXTENT SHOULD WE EXPLORE THE UNIVERSE AND WHAT IS THE CURRENT STATE OF KNOWLEDGE OF OUR SOLAR SYSTEM?

Exploring the universe provides opportunities to understand fundamental questions about nature, harvest natural resources, and possibly find suitable places for colonization. The extent to which we continue to explore for the sake of advancing knowledge is an open question—but one that demands consideration. Research-based exploration often spawns other technologies; thus, these efforts are valuable. Future endeavors need both public and private sources of funding due to the costliness of space travel and exploration. Already, companies like SpaceX design, build, and launch advanced rockets and spacecraft. SpaceX began its commercial missions for private customers in 2013.

Harvesting valuable materials from space is an intriguing endeavor, and possibly lucrative if these efforts prove feasible. To determine suitable sites for mining in space first requires explorers to identify likely sources. From remote and close-up studies, asteroids consist of numerous minerals. S-type asteroids contain metals such as nickel, cobalt, gold, platinum, and rhodium. A small 10-m asteroid has about 50 kg of platinum and gold. At current (2024) prices, 1.0 kg of gold is worth \$84,481 and 1.0 kg of platinum is worth \$30,890.

Colonization of space has the potential to find new places for humans to live or launching points for future exploration. Two of the most likely candidates for colonization are the Moon and Mars. The Moon is closer to Earth, but conditions there are harsh, and resources are limited. Mars also has harsh conditions but is a better choice for colonization due to the availability of more resources as compared with the Moon. Mars has carbon, nitrogen, hydrogen, and oxygen, most notably in the form of carbon dioxide, nitrogen gas, and water ice. All these materials and numerous compounds formed from them are useful for human habitation.

Models of the origins of our solar system are complex, incomplete, and ever-evolving; however, all viable ones consider certain facts. Known facts regarding the solar system's motions and organization reveal several trends. Planetary orbits are nearly circular and lie in approximately the same plane. Rotationally, planetary motions about the Sun, planetary motions (spin) about their axes, and satellite motions all mimic the Sun's spin about its axis. Besides motional characteristics, the structure of the solar system reveals planets, isolated in space relative to one another with high-density planets closer to the Sun and lower-density ones farther away. Other bodies within the solar system such as asteroids and comets contain primitive materials whose presence is consistent with proposed models.

These known facts suggest formation from a single event, occurring some 4.6 billion years ago. The uniformity of motions and accumulation of planets at nearly the same time rule out a buildup of planets and moons from separate events. The condensation theory of solar system formation begins with a rotating cloud of gas and dust. As the cloud contracts due to gravitational forces between particles, the system spins faster as predicted by conservation of angular momentum for a rotating body. (This effect is analogous to a figure skater rotating more quickly as they move their arms closer to their axis of rotation.) As the system contracts and spins more quickly, particles within the cloud concentrate in a plane. This accumulation of matter in a plane provides the conditions from which circular planetary orbits with a preferred direction of rotation emerge.

Dust particles contribute to the condensation of matter into larger masses by radiating energy and thus cooling the rotating matter and by serving as condensation nuclei. These nuclei act as microscopic platforms to which additional matter attaches to form larger and larger masses.

Cooling of matter within the rotating disk is crucial so that gas pressure does not overcome gravitational effects, thus preventing further accumulation of matter.

WHAT ARE BLACK HOLES?

As indicated by their name, black holes are regions of space–time that are dark to the rest of the universe. Black holes typically form when a high-mass star remnant collapses into a singularity and becomes unobservable by direct observation. Gravitational effects within a certain radius—known as the Schwarzschild radius—around the black hole prevent even the escape of light.

An imaginary sphere with a radius equal to the Schwarzschild radius defines a surface called the event horizon. No event within the horizon is accessible to an observer outside that surface. The Schwarzschild radius depends on the mass of the collapsed core and has typical distances of a few kilometers. For example, a star whose mass is comparable to our Sun has a Schwarzschild radius of 3 km.

One of the indirect ways to observe black holes is by gravitational lensing. This effect occurs when a black hole lies between a source and an observer. As light passes near the black hole, light rays are bent by the extreme gravity associated with the black hole. In addition, light from the background star brightens due to the intense gravitational field of the black hole. Instruments such as the Hubble Space Telescope look for distortions in starlight as the black hole drifts in front of background stars.

WHAT IS DARK MATTER?

One of the intriguing questions in modern astronomy is the so-called missing mass problem. In the 1930s, Swiss astronomer Fritz Zwicky noticed a discrepancy while studying the Coma Galaxy, containing more than 1,000 individual galaxies. Zwicky's observations showed that galaxies within the cluster move faster than predicted by their size and observable mass. To account for these effects, Zwicky suggested that additional mass is present, hence the reference to dark matter.

In the 1970s, Vera Rubin and colleagues confirmed the existence of dark matter by studying galaxy rotations. They observed that rates of rotation of galaxies do not decline with distance from the core as predicted, thus confirming that extra mass is present. Such matter is not detectable by usual techniques but is necessary to account for the observed rates of rotation.

Today, scientists acknowledge existence of this new form of matter that is unlike all other known particles or collections of particles. Current models depict dark matter as a web-like structure that permeates the entire galaxy. Dark matter occupies space and has mass but does not reflect, absorb, or radiate light to any detectable extent. Some theories assert that dark matter consists of unidentified particles that rarely interact with normal matter. One effort focuses on studying residual radiation from the Big Bang (known as Cosmic Background Radiation) to determine information about the distribution of dark matter.

CHAPTER 8

Electronics in the 21st Century

M ODERN ELECTRONICS ORIGINATED WITH the work of Lee DeForest, who in 1906 observed changes in vacuum tube currents due to small, variable electrical signals on the tube's grid. The concept of one signal controlling another led to the development of the first amplifiers in modern electronics and is central to many modern transistor circuits today. The progression of electronics continued with the invention of the point contact transistor by scientists, William Shockley, John Bardeen, and Walter Brattain, at Bell Laboratories. They received a Nobel Prize in Physics for their contributions to this invention in 1956. (Bardeen also shared a second Nobel Prize in 1972 for his contributions to superconductivity.) A third revolution in modern electronics followed in 1959 with the invention of the integrated circuit (IC) by Jack Kilby. ICs contain numerous transistors and other circuits within a single package. Arrays of ICs connected on a circuit board form the heart of many modern devices.

ATOMIC STRUCTURE, SEMICONDUCTORS, AND DOPING

The atomic structure of materials determines their electrical properties, which, when controlled and tuned, provide scientists and engineers the required electronic configurations to build devices. Negatively charged electrons within atoms possess potential energies determined by their

DOI: 10.1201/9781003511472-8

interactions within the system. As a result, electrons occupy specified energy levels or shells, roughly corresponding to certain distances from the atomic nucleus. Due to limitations imposed by quantum mechanics, each shell accommodates only a certain number of electrons. Outer shell electrons experience weaker attractive forces compared with inner shell electrons due to larger separations from the positively charged nuclei. (In addition, inner electrons screen the outer ones, further reducing attractive forces exerted on outer shell electrons.)

Valence electrons reside within the outermost electronic shells (orbits). The valence band refers to the range of energies in which the valence electrons reside. Valence electrons usually participate in bonding between atoms, but in some cases, gain sufficient energy to become unbound and part of a sea of so-called conduction electrons. Conduction electrons are free to undergo motions that produce flow of charge in a circuit (electrical current). The range of energies in which the conduction electrons reside is the conduction band.

Three general classes of materials referred to as conductors, semiconductors, and insulators comprise nearly all electronic circuits. Insulators are materials that do not readily conduct charged particles (generally electrons) due to large energy separations between their valence and conduction bands. Essentially too much energy is necessary to free the valence electrons to produce conduction. Semiconductors are materials with valence and conduction bands separated by small gaps so that valence electrons easily transition to the conduction band. Conductors are materials in which the valence and conduction bands overlap so that some valence electrons already have enough energy to participate in conduction.

Covalent bonds arise when atoms share valence electrons. During synthesis of diamond, the four valence electrons of carbon form four covalent bonds with neighboring atoms. As a result, each carbon atom effectively has eight valence electrons, thus producing a stable three-dimensional lattice structure. (Silicon has a similar outer electron structure as that of carbon.) Modifications to this stable bonding structure often provide the needed charge carriers to produce electrical current.

Control of conductivity in semiconductors occurs via doping. Doping is the process of incorporating impurity atoms to replace a certain fraction of silicon atoms in bulk silicon. These impurities introduce either an excess of electrons or holes into the lattice, thus providing (mobile) carriers that are available for conduction.

When atoms such as arsenic, phosphorus, and antimony with excess valence electrons substitute for silicon atoms, four of their outer electrons participate in bonding, leaving one free electron per impurity atom. Extra electrons within these n-type materials contribute to the conduction band and serve as the majority carriers in these materials. Conduction in n-type materials occurs due to motions of electron majority carriers.

When atoms such as aluminum, boron, and gallium with deficiencies of electrons substitute for silicon atoms, three of their outer electrons participate in bonding, leaving one hole per impurity atom. Deficiencies of electrons (holes) in p-type materials leave vacancies in the valence band and serve as majority carriers in these materials. Conduction in p-type materials occurs due to motions of hole majority carriers.

Flow of electrons or holes within the silicon lattice—doped intentionally to generate desired electrical properties—produces electrical current. Free electrons move under the action of a voltage applied to a conductor (or semiconductor). These motions result in transport of electrons through the circuit to form electron current. Hole currents arise in the valence band due to electrons attached to atoms moving to nearby holes with little change in energy. Once a valence electron transitions to a hole, another hole forms in a neighboring atom. Thus, the holes move from one site to the next, leading to the transport of holes and thereby a hole current.

PN JUNCTIONS, TRANSISTORS, AMPLIFIERS, AND ICS

A bulk piece of silicon with n-type dopants on one side and p-type dopants on the other, forms what is known as a pn junction. Once formed, excess electrons from the n-doped material diffuse across the junction and combine with holes on the p-side. As electrons diffuse, they leave behind atoms with excess positive charge, thus the n-region near the junction becomes positively charged. By analogous arguments, the p-region near the junction becomes negatively charged.

Devices with single pn junctions, referred to as diodes, have properties uniquely suited for various applications. Ideal diodes conduct when positive voltage appears on the p-side of the junction and negative (or zero) voltage appears on the n-side. This application of voltage is known as forward bias. Reverse bias refers to application of voltage in the opposite direction, thus reducing the current to zero. The ability of diodes to switch between these two states depending upon the applied voltage makes them ideal for building devices. One such example is an arrangement of four diodes known as a full-wave bridge rectifier, which converts AC (alternating current) into DC (direct current).

After conversion from AC to DC, the DC voltage has a fixed sign but pulsates in time. Capacitors added to these circuits change the pulsating DC into a constant voltage. DC is important for charging batteries and operating many electronic circuits while AC voltage is useful for running most devices in homes and businesses. Power companies produce AC due to the ease of transmission and the ability to deliver it at various voltage levels.

While single-junction devices like diodes are useful for AC to DC conversion (rectification), devices with two "semiconducting" junctions are ideal for controlling currents and amplifying AC input signals. One of the most common devices is the npn bipolar junction transistor. This transistor consists of three doped regions referred to as the emitter, base, and collector. In the npn version, the collector and emitter are n-type regions, and the base is a p-type region. Emitter electrons reach the collector under application of appropriate collector and emitter voltages and the presence of base current (Figure 8.1).

When the transistor operates with a negative (or zero) voltage applied to the emitter, free electrons enter the base region. Given the small length of the base region, most electrons pass through the base region without recombining with holes, move into the collector region, and then produce collector current when the collector has a suitable voltage. A small fraction of the electrons entering the base combine with holes there so that a small (base) current leaves the base region when the base has a proper voltage (biasing).

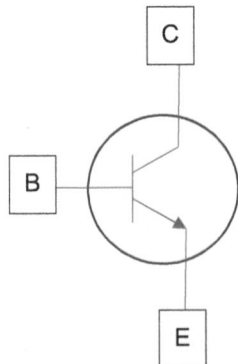

FIGURE 8.1   Bipolar junction transistor with collector C, base B, and emitter E

To generate the base current required to produce electron flow from the emitter to collector (collector current), the base requires a positive voltage. When both the base voltage and collector voltage are positive, a small fraction of the electrons enteringthe base undergo recombination to produce a nonzero base current. A larger fraction of electrons from the emitter pass through the base and arrive at the collector. This (collector) current is available for operating loads (devices) connected in series with the collector. Setting the base voltage to zero (or slightly negative), reduces the base current and therefore the collector current to zero. In this state, any device connected to the collector side of the circuit transitions to the "off" state.

Another crucial application of transistors is amplification. Amplification refers to increasing the magnitude of an electrical signal via specialized circuitry. Voltage amplification occurs by feeding an AC signal into the base of a bipolar junction transistor. During the half-cycle when the AC voltage increases positively, the base current increases, thus increasing the collector current. As the collector current increases, the collector voltage decreases below its usual voltage set by a DC power supply. This decrease occurs because a resistor is placed between the collector power supply and the collector itself. The decrease in voltage at the collector is magnified due to large collector currents developed in comparison with the base currents. Thus, small swings of the base voltage produce larger swings of the collector voltage. Under these conditions in which the output voltage is much larger than the input voltage, the transistor behaves as an amplifier.

Multiple transistors and other electronic components, interconnected on a single device, form an IC. Construction of ICs generally involves etching several components onto a single piece of silicon to maximize the number of circuit elements within a given area. Many electronic devices, including computers, smartphones, and televisions incorporate ICs into their designs.

Optical circuits have the potential to revolutionize modern electronics, although everyday applications are not yet widely available. Like their electrical counterparts, optical transistors act as switches or amplifiers using only light to control light. In current-day optical communications, electronic circuits control signal routing. These processes require electronic-to-optical conversions, which reduce efficiencies. In principle, all-optical systems of transmission and switching are faster and more efficient than electronic or optical-electronic systems.

COMPUTERS

The internal workings of a modern computer require processing, data transmission, storage, and connections to outside devices. The central processing unit (CPU) performs most of the processing by running programs and/or executing instructions. The CPU relies on memory units, given that some commands require analysis of data already available (or data produced during execution). Several units within the CPU perform their own specific tasks. These units consist of the arithmetic logic unit (ALU), the instruction decoder, the register array, and the control unit.

The ALU is the main CPU processing component, handling all arithmetic operations and logic operations. The ALU receives instructions from the control unit. The instruction decoder converts instructions into appropriate commands for the ALU. During operation of programs, the register array stores data and memory addresses. The availability of these data allows the programs to run efficiently. The control unit executes decoded instructions. The control unit provides timing, control signals for importing and exporting data, and synchronization for execution of instructions.

Data transmission in computers occurs along pathways known as buses, consisting of both physical connections and electrical specifications needed to transport signals. Serial and parallel buses send data either one bit at a time (serial bus) or simultaneously via multiple pathways (parallel bus). A common example of a serial bus is the USB or Universal Serial Bus. Three internal buses—known as the address bus, data bus, and control bus—provide interconnections between the CPU and memory and storage. The address bus carries information that selects ports and specifies memory locations or addresses. The data bus transfers program instructions and data between the CPU, memories, and ports. The control bus sends control signals to and from the CPU.

Computer software consists of two main categories known as system software and applications software. System software (often referred to as the operating system) provides the interface between the user and the computer. Applications software refers to written codes that accomplish certain tasks such as word processing. Many occupations use specific applications software to make workers' lives less tedious. For example, accounting software performs calculations and prepares tables, graphs, and reports as requested by the user.

AI refers to devices or systems that perceive the world around them and apply learning to develop appropriate responses to defined goals. Today

AI is instrumental in performing online searches, answering interactive questions, and delivering online options based on our habits. Applications of AI to fundamental research include controlling optical set-ups to achieve conditions needed for exotic experiments and searching for patterns in astronomical data to identify and/or validate the existence of exoplanets. In the future AI has the potential to control movements of disabled persons or record their thoughts if they are unable to speak.

WHAT ARE THE PRIMARY ADVANTAGES OF OPTICAL CIRCUITS?

The primary advantage of optical circuits is the high speed with which light travels in optical circuits. Light in vacuum moves at a speed $c = 3.00 \times 10^8$ m/s, whereas light in a fiber optic cable moves at approximately $0.67c$ or 2.00×10^8 m/s. At these speeds, light covers 2 mm within a fiber in 0.00000000001 s. In traditional electrical circuits, individual electrons travel at speeds of approximately 0.02 cm per second, thus particles traverse 2 mm in about 10 s.

While drift speeds in traditional circuits are quite slow, overall response times of electrical circuits are much faster given that electrons within conductors interact with each other. As a single electron moves, it influences other electrons, and those, in turn, influence additional ones along the conducting path. The overall result is that electrical circuits respond quickly when connected to a voltage source. Upon application of a voltage, electrons migrate from site to site along the conducting path, thus producing a continuous flow of charge (current) throughout the circuit.

ARE TRADITIONAL ELECTRONIC DEVICES SUSCEPTIBLE TO NUCLEAR ATTACKS?

Effects of space radiation on electronic devices to determine their viability beyond the confines of Earth is the focus of extensive studies. Such studies also inform designers how well devices withstand nefarious radiation, produced by detonation of nuclear (fission) bombs. When nuclear detonation occurs, radioactive dust and ash fall from the sky. Fallout materials usually consist of fission products, along with nearby atmospheric atoms that have absorbed neutrons to become radioactive.

Radiation damage to electronics occurs due to single events or multiple ones over time. Total dose effects refer to changes in electrical properties from incoming radiation due to accumulation of damage. Such changes over time cause electronic devices to drift outside their design parameters, resulting in operational failures. Another cumulative effect is displacement damage resulting from incoming particles causing displacement of silicon atoms. As additional displacement occurs, points of low electric potential (voltage) develop within the lattice, thus trapping electrons. These effects cause the electrical conductivity of damaged devices to decrease, thus impacting their abilities to function properly.

Several single-event effects also lead to immediate damage (or immediate alterations) to electronic devices. A single-event upset causes an unplanned change in the logic state of a digital electronic device. Such events are temporary, thus restoration of the original (intended) state is possible without replacing devices. Single-event burnouts, however, result in destruction of the device.

Radiation damage is a serious threat to modern electronic devices due to various scenarios, ranging from ones that cause destruction of devices to those that cause failure due to long-term effects.

WILL AI REPLACE HUMAN DECISION-MAKING?

The prospect of losing our abilities to make decisions and shape our futures due to the widespread use of AI is bewildering to many in the public. Several years ago, when AI gained enough proficiency to beat the chess grandmasters, reactions revealed both disappointment and excitement. Many expressed disappointment that humans no longer prevailed over human-developed machines. Others felt excitement to know that machines could teach us how to play chess, along with helping us cure cancer and solve our sustainable energy crisis.

The ability to play games more effectively has wide-spread applications—including how to defeat diseases. In games designed to model disease, physicians assume the role of predator and cancer cells become the prey. The physician proposes therapies to eliminate cancer cells but also allows them opportunities to respond by evolving therapy resistance. The physician, in turn, varies treatments in response to how the cells behave. The winner of the game is the one who achieves certain prescribed objectives.

AI is already prevalent in many parts of our lives. For example, massive computer search engines use AI to provide data on demand. Most folks are very comfortable getting information quickly and efficiently but are less comfortable with machines keeping records of our searches and using that information to personalize online ads. In former days, salespeople observed shoppers and noted what items piqued their interest. Salespeople then approached customers and said, "I noticed you browsing our dining room tables. Would you like to see others in another section of the store?" The concept of noting our habits and responding is similar in both cases, but the machine watching us seems more intrusive and threatening.

Another emerging trend is online customer service supported by AI. The "support AI" takes questions and responds with appropriate answers, based on loads of available data. Thus, we get accurate answers with little delay. Given the explosion of AI applications, some folks now converse with AI devices as part of their social or even romantic interactions.

Ultimately, the most effective use of AI promotes human health, welfare, and security while controlling energy usage and limiting demands on our ecosystem. Admittedly this is a tricky balance. We find ourselves on a planet that already exceeds its human carrying-capacity by three times, while petroleum remains our primary energy source. To thrive over the next 100 years, we must continue to innovate and creatively implement technology. Thus, we will need a comprehensive strategy and an ethical framework to guide decision-making on AI initiatives.

CHAPTER 9

Our Energy Future

FOR ALL LAND-BASED ORGANISMS on Earth, the primary source of energy is the Sun. Electromagnetic waves (light) from the Sun initiate all photosynthetic reactions in plants to produce various sugars and cellulose. Plants provide nourishment for most of the animal kingdom, including humans. Even the decay of plant life under extreme temperatures and pressures over millions of years leads to the formation of oil, coal, and other fossil fuels such as natural gas.

In addition to consuming foods for nourishment, humans require energy to keep our bodies at comfortable temperatures and to transport people and goods from one location to another. On larger scales, energy is necessary to produce manufactured products, generate electricity to operate a myriad of devices, and develop a robust infrastructure for a modern society.

One of the emerging questions of the 21st century is how to obtain the energy needed to sustain our species. Given dwindling resources and a growing population, solutions require improvements in efficiencies at all levels of the energy economy, a larger repertoire of energy sources and storage devices, and conservation efforts to avoid using resources too rapidly. In addition, re-evaluation of energy usage in how we produce food, how we transport ourselves (and how often), and what entertainment activities we pursue is critical to our efforts.

DOI: 10.1201/9781003511472-9

ENERGY AND WORK

Energy is a subtle concept, and one that is difficult to define. In the physical world, energy generally refers to an ability to cause motion, and, more specifically, to do work. Work done implies that a mass (body) undergoes a change in its position (called displacement) due to an applied net force. For purposes here, a force is any kind of interaction between objects that results in a push or pull.

To clarify the concept of work, an example is useful. Suppose you want to move a book across a smooth, flat tabletop. The easiest approach is to push the book until it is in a desired location. This example of pushing the book (applying a force) requires an amount of work that depends directly on both the force applied and the change in position of the book. In mathematical terms, the work is the product of the force applied in the direction of the motion and the amount by which the position changes.

What is the relationship between work and energy? Doing work requires expenditure of energy. Motion of a book, initiated by a human force, requires contraction of muscles due to chemical reactions. Contraction initially involves formation of bonds between thin and thick muscle filaments. In a muscle fiber, thin and thick filaments alternate so that they form an interleaved pattern. Once bonds are formed between the filaments, such bonds are activated by breakdown of available ATP. Bonds then undergo bending, thus pulling sections of muscle filaments together to cause contraction. Muscular contractions generate forces needed to perform work.

A nonhuman example of work and energy is the operation of a turbine-generator system in a power plant. In a power plant, some source such as coal, natural gas, or nuclear fuel releases energy, absorbed by a medium like water to form steam. Once under enough pressure, the steam turns a turbine—a device that rotates when a fluid moves through it. The unit for energy is the Joule (J). For reference, a human of mass 70 kg moving at a speed of 4.0 m/s possesses 560 J of kinetic energy. One of the unique features of energy is that it can exist in various forms, often characterized by their methods of storage or production.

Mechanical energy originates from motions of systems or configurations of systems. (For example, a book resting on a table has the potential to move if it were pushed to the edge.) Energy of motion is kinetic energy, whereas energy associated with a system's location or configuration is potential energy. Possessing kinetic energy is useful for transporting people and supplies, whereas potential energy is useful for storing energy.

For example, water pumped to the top of a hill has potential energy capable of performing work when needed.

Certain kinds of chemical reactions, such as oxidation (burning) of wood, release (chemical) energy. (In some cases, energy is necessary to initiate reactions.) Energy from the Sun along with the presence of chemicals such as carbon dioxide and water produce plant materials and oxygen during photosynthesis in green plants. In the human body the breakdown of various energy sources produces motions of the body or maintains bodily functions needed for life.

Electrical energy involves stored charge or charge in motion. (Charge is the fundamental quantity used to describe static cling of clothes in a dryer or explain the discharge of lightning.) Motions of charges are common to all devices that operate using electricity. Power plants produce large quantities of electrical energy to operate whole cities while nerve cells produce small electrical impulses to transmit signals from one nerve cell to another.

Emission of EM waves, ranging from long-wavelength radio waves to short-wavelength gamma rays, produces radiant energy. Emission occurs due to several atomic-scale processes. Generation of visible light occurs by first exciting electrons within atoms (or molecules) to higher energy levels via heating or electrical methods. When excited electrons decay back to lower energy levels, they emit energy. Radiant energy is responsible for providing input energy for solar panels, generating the feeling of warmth on your face from a campfire, and transmitting signals to your cell phone.

Changes within the nucleus of atoms release nuclear energy due to conversion of mass into energy. During fission a parent nucleus separates into two smaller (daughter) nuclei with a concomitant release of energy. Nuclear fission is the principle way present-day reactors work to generate steam to drive turbine-generators. By contrast, during nuclear fusion smaller nuclei join to form larger ones such as what happens inside stars. (In ordinary stars, hydrogen nuclei fuse to form helium nuclei.)

POWER

Another way to characterize energy expenditure (or delivery) is to specify how quickly a system consumes (or delivers) energy. This ratio of the work done, or energy transferred, per unit of time (usually 1 s) is a measure of power. The unit of power is the Watt, named after James Watt, a Scottish inventor whose insights improved the efficiency of the original steam

TABLE 9.1 Power ratings of various systems

System	Power (Watts)
Sun	10^{26} W
Power plant	100s MW to a few GW
Car	75 kW
Elite sprinter	2,000 W
Solar panel	100–500 W
High-efficiency light bulb	15 W

engine. For reference, 1 Watt is the approximate power to move a soccer ball from floor level to a height of 0.24 m in a single second (Table 9.1).

From a known power, the amount of energy required (or generated) is the product of the power and an appropriate time interval. For example, an efficient LED light operating at 15.0 W for 4.0 h (14,400 s) requires 2.16 × 10^5 Joules of energy. In electrical applications, the product of the power of a device in kilowatts (kW) and the time of operation in hours gives the energy usage in kilowatt-hours (kWh)—a quantity used to compute how much we owe for our electricity bills.

CONSERVATION OF ENERGY

Energy is present in a variety of forms but is constant within a closed universe. This statement is known as the law of conservation of energy and is consistent with all experimental results. One of the subtleties of energy conservation is the ability of energy to become matter and vice versa. Thus, careful analyses are crucial for examining energy changes in practical devices and processes.

A power plant is a system that takes energy from a source (such as oil, coal, natural gas, or even nuclear fuel) and transforms a portion of that energy into electrical energy. Burning of a combustible fuel inside a boiler releases energy that converts water to steam. This conversion raises the thermal energy of the water. Once the water is at sufficient temperature and pressure, it flows through nozzles onto the blades of a turbine, causing the turbine shaft to turn. (Here chemical energy from the source changes into mechanical (rotational) energy during this process.)

Rotation of the turbine shaft, connecting the turbine to an electrical generator, turns the generator. An electrical generator typically consists of a coil of many turns of wire that rotates within a magnetic field. This action produces an electrical voltage that depends on several factors, including the rotation rate of the coil, its area, the number of turns of wire, and the strength of the magnetic field.

After the steam passes over the blades of the turbine, it passes through a condenser where it converts back to a liquid (and has the same amount of thermal energy it did initially). The condenser operates by taking water from a lake or river and passing it through the coils in a heat exchanger. The exchanger is in contact with the used steam to promote condensation. Cold water pumped from the lake experiences a temperature increase during this process, thus creating some thermal pollution. The condensed water is pumped back to the boiler unit, and the process begins again.

Power plants operate in a cycle that extracts energy from a source and converts that energy into another useful form (typically electrical energy). For purposes of conservation, inputs of energy are primarily from the fuel, air, and thermal energy of the cooling water for the condenser. Outputs of energy come from the electrical energy produced by the power plant, the thermal energy of the hot (warm) water leaving the condenser, and the energy of the hot combustion gases leaving the stack. In equation form, conservation of energy for a power plant is:

$$E_{fuel} + E_{air} + E_{water\ in} = E_{electrical\ energy} + E_{water\ out} + E_{combustion\ gases}.$$

One of the measures of any practical device is its efficiency. Efficiency refers to the amount of useful work (or energy) produced relative to the total energy input. A device that is 100% efficient converts all its energy input into useful work. Knowing efficiencies is crucial for all design work. For example, power plant engineers must know the amount of energy required to produce a specified amount of electrical energy. Increases in efficiencies reduce the amount of fuel input needed, thus saving (input) energy costs.

Many processes in nature involve converting energy from one form to another. Energy needed to produce motions of the human body comes from chemicals within our muscles, along with others stored in organs like the liver. Chemicals needed for the body originate in foods containing basic compounds such as carbohydrates, fats, and proteins. Plant-based foods have energy-containing compounds produced via photosynthesis. Each step in the process of generating motion involves energy conversion and has a particular efficiency. The overall efficiency is the product of the efficiencies involved in each step.

In another example, a power plant extracts energy from a fuel, converts a portion of that energy into electrical energy, transmits that energy via power lines, and then delivers that energy to a lighting device. The overall efficiency of this process is:

$$E = E_{chemical\ to\ electrical} E_{transmission} E_{electrical\ to\ light}.$$

A typical efficiency for this process is 1.6%, given an efficiency for chemical-to-electrical conversion of 35%, an efficiency for electrical transmission of 90%, and an efficiency for electrical-to-light conversion of 5% for traditional (incandescent) lighting.

ENERGY AND THE HUMAN BODY

As humans, we consume energy to keep our bodies functioning, maintain comfortable temperatures in our homes, and satisfy our desires for play and entertainment. As a living machine, the human body typically consumes energy at a rate of 100 J during each second of our adult lives. This rate of energy consumption, known as the basal metabolic rate (BMR), is known from studies of human metabolism and assessments of nutritional requirements.

From this rate of consumption, the number of food calories needed results from several simple conversions. If BMR is constant over the course of a day, the body uses about 100 J each second over 24 h (86,400 s), for a total of 8.64×10^6 J. Each calorie is equivalent to 4,186 J, so, over the course of a day, a typical BMR translates to 2,064 Calories. (This value represents a typical caloric intake; however, particularly active folks require more.)

In the developed world, most people have sufficient food to consume at least 2,000 calories per day, but others consume far less. To gain perspective on human food needs, let us assume that 8 billion people worldwide have a caloric intake of 1,200 calories each, translating to 4.0×10^{16} J per day. The Sun delivers approximately 1,000 J per second per square meter at the surface of Earth in full, direct sunlight. To deliver an equivalent amount of solar energy over 24 h of full sun requires an area of 4.6×10^8 m². This value does not consider the land area needed to produce food—only the land area needed to deliver an equivalent amount of solar energy.

The reality is that the area cited underestimates the land (area) needed to produce food because not all incident sunlight converts into usable foodstuff. If rice is the primary source of calories in the calculations above, a total of 7.4×10^9 kg of this food is necessary each day for the entire world-wide population. Each kilogram of rice produced requires a land area of 2.8 m^2, so the total land area needed to produce food for all persons on Earth for one day is 2.07×10^{10} m^2. Such an area is equivalent to four times the land area of the state of Delaware.

One of the overriding concerns of those who study the sustainability of our planet is the so-called carrying capacity. Carrying capacity refers to the number of people that a given environment can support sustainably over time. Current estimates suggest that the Earth's carrying capacity is between 2 and 4 billion people while the world population is over 8 billion. This disparity indicates that our current population trends are not sustainable given current energy demands.

ENERGY SOURCES

Beyond fuel for the body, modern societies rely on energy sources to produce electricity for households, run transportation systems, and make the products that we consume. Many energy sources are available, including fossil fuels, nuclear materials, and renewables such as wind and solar. Specialized energy sources include geothermal. For reference, details of nuclear energy are covered extensively in Chapter 6.

Petroleum is a fossil fuel that forms from the decay of organic material, particularly marine life. To produce petroleum, decaying matter undergoes changes under extreme pressures and temperatures. Over the course of millions of years, dead matter converts to gaseous or liquid hydrocarbons that migrate through rocks and structures below ground to form deposits. Various kinds of rocks, including shale, limestone, and sandstone, act as sponges to retain oil deposits within them. Trapping of oil between impermeable or nonporous rock above and below oil-forming areas causes petroleum deposits to accumulate. Left undisturbed these pools of oil remain for millions of years. Petroleum serves as the starting material for gas and diesel fuels, along with many plastic products.

Another fossil fuel, known as natural gas, consists of methane (CH_4) and other lightweight hydrocarbons. Natural gas forms from decayed organic matter and occurs either alone in reservoirs (nonassociated gas) or in the same reservoir as crude oil (associated gas). In 1821, workers completed digging of the first US natural gas well in Fredonia, NY. To distribute gas to homes and businesses, workers constructed a natural gas pipeline network following World War II. In the United States the main use of natural gas is for home and water heating and cooking. With the advent of fracking, many oil fields such as the Bakken field in North Dakota now produce extraordinary amounts of natural gas.

Wind consists of moving air molecules, possessing kinetic energy, capable of converting mechanical motions into other forms of energy or performing work. Winds occur due to atmospheric pressure differences, driven partly by the input of solar energy. Harnessing of wind to accomplish useful tasks such as pumping water and grinding grain originated several thousand years ago in Babylon and China. Windmills gained popularity in Europe beginning in the 12th century. In the United States farmers used energy from windmills to pump water from the ground.

Today large-scale wind turbines produce sizable amounts of electrical energy in China, the United States, and Germany. Wind turbines work by converting the wind's kinetic (mechanical) energy into electrical energy, produced by rotating turbine/generator systems. Current land-based turbines have power ratings of 1–3 MW each. In March 2024, China's wind farms produced over 100 terawatt-hours (TWh) of electricity, the highest monthly total ever by a single country. Countries with the largest installed wind capacity include China with 475,000 MW, the United States with 150,000 MW, and Germany with 69,000 MW.

On a yearly average, the Sun delivers 342 J of energy per second to each square meter of the Earth's surface, equivalent to about 10,000 times humanity's current energy usage.. While solar energy is available across the globe, it requires transformation to produce electricity or provide heating.

Most technology today focuses on using sunlight to transfer energy to water or other absorbers or to produce electricity. A discussion of solar heating is in the text below. Electrical power plants also use absorbed solar energy to produce steam to drive electrical turbine/generator systems. A more direct way to produce electrical current is via solar (photovoltaic) panels, constructed of cells of doped silicon. As of 2022, the United States had 142 GW of installed solar photovoltaic capacity.

Part of each solar cell has an excess of electrons (n-doping), and the other part has a deficiency of electrons (p-doping). When light (contained within packets called photons) strikes the junction between the two regions, electrons from a lower band of energies (called the valence band) jump to an upper band of energies (called the conduction band). Once in the higher band, a voltage created across the junction causes the elevated electrons to flow in a circuit.

Water flowing naturally along a waterway or through a dam is another specialized source of energy. As with wind turbines the flow of a fluid (water) drives a turbine/generator system to produce electricity. Electrical power production ranges from several kilowatts for generators installed in rivers to 2,000 MW for the generators installed at Hoover Dam. In some locations, thermal energy from deep within the Earth is available as steam coming to the surface via natural processes. Alternatively, systems built to tap stored energy within the Earth harness steam or extract thermal energy.

AIR POLLUTION

A significant drawback of liberating energy from sources such as fossil fuels is pollution. Pollution of air comes in a variety of forms, ranging from floating particles to foreign gases in the air. Carbon dioxide as a pollutant contributes to climate change due to its role as a greenhouse gas, as discussed in Chapter 2. Other gaseous pollutants include carbon monoxide, sulfur oxides, and nitrogen oxides.

Incomplete combustion of gasoline fuel produces carbon monoxide (CO), a colorless, odorless, and poisonous gas. Carbon monoxide enters the bloodstream via absorption in the lungs. Once in the bloodstream, CO binds to hemoglobin, thus preventing it from carrying O_2 to the body's cells. As a result, the body suffers from oxygen deprivation symptoms such as dizziness, headaches, and visual changes.

Sulfur is another contributor to air pollution, especially in the form of oxides such as SO_2 and SO_3. These compounds form when sulfur, primarily released from burning fossil fuels, reacts with oxygen. Sulfur dioxide is harmful to humans, vegetation, and materials. In humans, sulfur dioxide causes damage to the upper respiratory tract and lung tissue and aggravation of existing lung disease. Several cases of extreme SO_2 air pollution have resulted in human deaths.

In 1948, 19 people died in Donora, PA, and in 1952, 4,000 people died in London. The concentration of SO_2 in London at that time was seven times its normal level.

As rain (or snow) falls through the atmosphere, water molecules encounter sulfur oxides and nitrogen oxides, producing precipitation that is acidic. This acid rain contains sulfuric acid and nitric acid that causes acidification of lakes, leading to significant declines in fish populations. Acidic precipitation also damages vegetation and causes corrosion of buildings and other structures.

Hydrogen ion concentrations determine the acidity of liquid solutions expressed on a pH scale. Each hydrogen ion has a deficiency of one electron; thus, a hydrogen ion has a single positive charge. A solution that has an equal number of positive and negative charges such as pure distilled water has a pH of 7. As pH decreases, the solution becomes more acidic; and as pH increases, the solution becomes more basic. Each unit change in pH corresponds to a factor of 10 change. For example, if pH decreases from 4.0 to 3.0, the solution becomes ten times more acidic.

Problems due to acid rain are particularly prevalent in the eastern United States, Canada, and northern Europe. Numerous lakes in New York's Adirondack Park have lost their native fish populations due to water acidification. Acidity also tends to cause corrosion and breakdown of infrastructure such as buildings, roads, and bridges, along with potential health effects for humans and other wildlife.

Other pollution problems arise due to the presence of very small (0.01 to 50 microns in diameter) particulates. Examples of particulates include dust, volcanic ash, pollen, sea salt spray, and fly ash. These so-called aerosols remain suspended in air over significant periods of time. Combustion processes typically produce particles that are less than 1.0 micron in diameter, whereas dust particles range from 1 to 1,000 microns. Particulates impact our ability to breathe, aggravate existing cardiovascular disease, and possibly damage the body's immune system.

HEAT PUMPS AND SOLAR (THERMAL) HEATING

While cooling during hot summer days is often desirable, most regions on Earth require at least some kind of heating for human survival. During pioneering days and before that time, the main source of heat was from combustion of firewood. (Even today burning of wood for heat is still in use, primarily in the underdeveloped world.) For most citizens, burning wood is prohibitive due to expensive fuel and the effort in maintaining and tending fires—not to mention increases in air pollution.

One convenient way to convert electrical energy into thermal energy for heating is via a heat pump. A heat pump is a device that works in a cycle to extract energy from a cold reservoir using a fluid as a storage medium, performs work on that medium, and delivers energy from that medium to a hot reservoir.

In heating mode an outside coil (carrying the working fluid) serves as the evaporator and absorbs energy from outside air (or other body) turning the fluid into a gas. The fluid then passes through a compressor that increases the pressure and thus the temperature of the gas. The working gas then enters the condenser coils, which are in contact with inside air. The warm gas transfers energy to the inside air, thus decreasing the temperature of the gas and causing it to condense (to a liquid). The liquid then passes through a pressure-reducing valve to the outdoor evaporator coil, and the process begins again.

Ratings for heat pumps examine a quantity known as the coefficient of performance (COP). COP is the ratio of the energy transferred to the hot reservoir to the amount of input work. Modern heat pumps have COPs in the range of 3.0–5.0. The COP of a heat pump decreases as the outside temperature drops (decreases) given that less energy is available in gases at lower temperatures. (The internal energy of a gas is proportional to its temperature.) To make extraction of energy more efficient, designers place the evaporator in contact with a reservoir such as the ground at a depth of a few meters or a body of water that has a nearly constant temperature.

The delivery of multiple units of energy per single unit of input energy (COPs > 1.0) suggests that heat pumps are efficient devices for keeping humans warm. These values, however, often are misleading, because heat pumps operate using electricity, produced by a primary source of energy such as fossil fuels, the Sun, or wind. Electrical energy generated from these sources has conversion rates that are 30% or less, thereby reducing the overall conversion rate.

In addition to indoor heating, many modern homes use devices to heat water. In the traditional design, natural gas or electricity supplies energy for heating. Once the heater is set to a desired temperature, it operates to establish that temperature and maintain it. The main disadvantage of these kinds of heaters is the amount of energy required to maintain the desired temperature of the water even if no usage occurs.

Given this drawback, a new generation of hot water heaters produces hot water on demand. When needed, water runs through a heat exchanger that has a natural gas burner or electric element. Such devices require high power but only during times in which hot water is needed. Overall, 40% reductions in energy bills are possible using these water heaters.

Instead of using fuel such as natural gas, certain houses employ passive solar designs for space heating. Here the house itself serves as the solar collector and storage medium. When properly constructed, natural flow of air within the house distributes energy to maintain warmth without the need for pumps or fans. Objects inside the house absorb sunlight entering the house from south-facing windows. Materials that absorb large amounts of energy include concrete, water, and stone given their large thermal masses.

To facilitate energy absorption in winter and prevent it in summer, passive solar houses usually have special roof overhangs. Requirements for a passive solar house include insulation, solar collection via south-facing windows, and thermal storage of energy inside the house.

Passive solar houses include three types: direct gain, indirect gain, and attached solar greenhouse. A direct gain system allows sunlight to enter the house via south-facing windows. Inside the house, thermal storage material such as a concrete floor (or walls) absorbs ambient solar energy. These massive systems then radiate absorbed energy at night, primarily as infrared radiation, to maintain comfortable temperatures. (Infrared radiation is the energy you feel from a campfire when you are several meters away.)

Indirect gain systems operate by absorbing energy in one part of the interior of the house and distributing that energy by convection and conduction. In one design, a massive concrete wall, located behind a large south-facing window, acts as the solar absorber. Wall construction provides space between the floor and the bottom of the wall and space between the top of the wall and the ceiling to promote convection. During the day as air between the glass and wall becomes warmer, it rises and passes through an upper vent into the adjoining space. Cool air near the floor from the adjoining space enters the glass-wall space through a lower vent so that warm air is continuously transferred to the living space.

At night the (upper and lower) vent controls close to prevent transfer of warm air via the reverse process. Warming of air inside the living space continues during nighttime hours, via radiation from the Trombe wall facing the living quarters and convection (also driven by the Trombe wall facing the interior) within the room. When lot space is available, some houses have an attached greenhouse that provides both a growing space for plants and a solar storage medium for household heating.

In conjunction with heating of the airspace inside a home, similar methods produce hot water for human use. A typical solar water heater consists of tubing inside a box with a transparent top and absorbing material on the bottom and sides. Light passing through the top of the box causes the interior temperature to rise due to absorption within the box. Water circulating through the tubing exchanges energy with the interior of the box, thus increasing the temperature of the water. Controls circulate water until it reaches a desired temperature.

THE FUTURE OF ENERGY

As a species, we are at a critical juncture as we seek to balance modern life with shrinking resources. Depletion of petroleum reserves will occur during the next few hundred years, given our current rate of consumption. Continued use of fossil fuels, diminishes air quality and leads to the build-up of CO_2 in the atmosphere and thus enhanced greenhouse effects. Rising global temperatures, in turn, increase the likelihood of events, such as severe weather, wildfires, and flooding. Given this situation, we need a comprehensive plan that addresses elimination of pollution, expansion of energy sources, specialization based on local resources, and conservation.

Production of electricity for the foreseeable future will include the use of fossil fuels and concomitant release of greenhouse gases. On this front we must develop and use the latest technology to control the amount of greenhouse gases released into the environment. Many strategies and technologies already exist to limit CO_2 release, so these must be employed whenever possible. As we transition to more and more renewable energy sources, release of CO_2 will decrease automatically.

To solve the problems of rising energy demand coupled with the release of CO_2 into the atmosphere, we must expand our repertoire of energy sources with an eye toward transitioning to more and more renewables. Solar and wind energy are renewable sources while nuclear energy is sustainable over thousands of years. In 2023, electricity derived from solar and wind made up 14% of the total generated in the United States. Electricity production from nuclear energy (19% of the total in 2023) will remain nearly constant given the lack of a national nuclear energy policy for nuclear waste storage. If nuclear fusion becomes feasible over the next decade or so, our renewable energy options could expand greatly. Other ways to effectively expand resources include pumping water to nearby hills for storage, generating hydrogen, or charging electrical storage devices during times in which excess electricity is available. These energy storage media serve as sources to produce electricity when demand is high.

Local resources often provide unique opportunities to produce energy. Geothermal installations in geologically active regions supply steam to produce electricity and/or generate hot water. Wind farms, installed in regions where wind is plentiful, produce enough electricity to supply whole regions of the country. Solar arrays, operating in regions with ample amounts of sunlight, also contribute to our renewable energy repertoire. As indicated, certain regions offer renewable energy options by virtue of their locations. However, renewable energy use has the potential to expand beyond these regions due to enhanced transmission of electrical energy over large distances.

Conservation efforts offer a range of scenarios. To start, experts recommend cutting extraneous usage of electricity at home, improving insulation of residences to conserve heating fuel, and developing energy-saving strategies for local travel in vehicles. More drastic changes impact lifestyle. Some folks now choose to live in areas where they can walk or bike to grocery stores, places of employment, and local entertainment districts. Moreover, their homes or apartments are smaller but have highly efficient spaces. Diets consist mostly of nonprocessed, plant-based foods, prepared just before consumption. (For reference, meat-based diets require several hundred times more land area than plant-based diets.) Controlling the number of calories consumed also saves energy given that each calorie of food requires a certain input of energy to produce. All conservation efforts and efficient lifestyle choices contribute to saving energy and reducing the effects of climate change.

DISCUSSION QUESTIONS

What Are Ways to Engage the Public in Energy Conservation and Advocacy?

Many in the public assume that energy sources are virtually limitless, and their use has little impact on our ability to sustain current lifestyles. The first step in engaging the public is to educate them to understand that both assumptions are incorrect. All energy sources buried within Earth, such as coal, petroleum, and natural gas, form on time scales of millions of years; therefore, these reserves have limited lifetimes. Current estimates of oil reserves suggest they will last a few hundred more years. In addition, all materials (coal, etc.) requiring combustion to release usable energy produce greenhouse gases and sulfur oxides and contribute to acid rain. Therefore, these traditional sources cannot meet our long-term needs.

Showing citizens that their efforts are impactful and that they are valuable participants in the energy economy also promotes engagement. Given the impacts discussed above, energy usage and policy affect each of us and, therefore, each of us has opportunities to contribute to solutions and conservation. Conservation on a global scale requires individuals to evaluate energy usage and respond with changes in habits, or more dramatically, changes in lifestyle. These efforts contribute to saving resources and helping families save money on energy.

Another way to empower citizens is through personal advocacy and action. Here folks engage their employers and local governments to enact energy-saving measures for buildings, equipment, and other facilities. Examples of actions include helping people conduct energy audits of their homes, eliminating food waste by connecting grocery stores with food agencies, and limiting local travel by walking and strategic use of vehicles. On larger scales, citizens exercise their power to lobby for stricter pollution controls, faster transition to renewable energy sources, and national policies to manage nuclear waste.

How Do We Ensure That Citizens Are Invested in Their Energy Futures?

Most citizens use energy in ways that are convenient. For most Americans, this translates to tapping into the electrical grid without much thought to monitoring usage, driving cars nearly everywhere we go, and eating highly processed foods. These actions do not encourage folks to consider how they might conserve energy or contribute to a sustainable energy economy. The section above discusses several engagement strategies while the text below proposes a model for investment in renewable sources.

Creation of incentives is the key to engagement of citizens. One way to incentivize our system for producing and distributing electricity is to make citizens of each state owners of all wind and solar power available within the state's borders. In this model, for-profit power (electric) companies pay a fee for each kWh of electricity produced from renewable sources. Funds generated are the property of citizens who exercise control over the revenue for various projects. Options here might include funding of state parks or public infrastructure (including an improved electrical grid), direct payments to citizens, and long-term investments that pay the citizen-owners lump sums at specified intervals.

Critics contend that ownership of renewable sources by citizens just raises electricity prices when power companies incur fees for using renewable resources. While this thinking is probably correct, citizen-users benefit in several ways. Investments in the electrical grid generally lower consumer costs over time due to higher efficiencies. Other projects such as parks and public spaces receive additional funding. Moreover, fees charged to consumers depend on electricity usage; therefore, consumers develop motivation to conserve electricity. Consumers also have options to reward power companies for lowering costs due to improvements.

Have Efficiencies of Devices Improved Recently?

The efficiencies of lighting devices have improved significantly over the past 20–30 years. Traditional lighting consists of evacuated bulbs containing filaments that become extremely hot when electrical current passes through them. The filaments emit a range of electromagnetic waves, including visible light. The drawback of traditional bulbs is that they only convert about 10% of input energy into usable light. Recent developments in fluorescent and LED lighting have increased efficiencies greatly.

Fluorescent bulbs produce light by exciting atoms within the bulb using a specialized electrical device. Once excited, these atoms emit ultraviolet (UV) light as they de-excite over time. The UV light strikes a phosphor coating within the tubes, causing those coatings to emit visible light. Efficiencies are better in fluorescent bulbs due to reduced temperatures and less conversion of energy into unwanted infrared radiation. Typical fluorescent bulbs require about 40 W to operate as opposed to 100 W for traditional lights that produce the same amount of visible light. LED lighting converts about 80% of input energy into usable light by passing current through semiconducting devices called diodes. As of 2020, 47% of US households report using LED lighting.

DISCUSSION QUESTIONS

1. Summarize your proposed energy policy.
2. How would you convince citizens to conserve energy?
3. Is energy waste a problem? Investigate.
4. Are you in favor of increasing electricity supplied by nuclear energy? Discuss.
5. Is our energy situation hopeless? Give an assessment.

Index

Note: Page numbers in **bold** refer to tables.

A

Acetylene, 43
Acid rain, 41, 98–99
Address bus, 86
Adrenalin, 45
Aerobic decomposition, 42
AI, 86–87, 88–89
Air pollution, 34, 98–99
Alcohol, 44
Aldehyde, 45
Alkaloids, 45
Alkynes, 43
Alpha particles/rays, 57, 65
Alternating current (AC)
    circuits, 49–51, 84
    generator, 52
    voltage, 84, 85
Amines, 45
Amplification, transistor application, 85
Amplitude, 12
Angioplasty, 29
Angiotensin-converting enzyme (ACE)
    inhibitors, 29
Application software, 86
Arithmetic logic unit (ALU), 86
Aromatic hydrocarbons, 43
Arrhythmia, *see* Ventricular fibrillation
Aspirin, 45
Asteroids, 77–78
Atoms
    atomic structure, 81–83
    atomic theory of matter, 4–5
    early models, 56–57
    fluorescent bulbs, 105
    interstellar space, 72–73
    organic molecules, 42–46
    water molecule, 39–40

B

Bardeen, John, 81
Basal metabolic rate (BMR), 95
Benzene, 43
Beta rays, 57
Big Bang, 71, 76, 80
Binding energy, nucleons, 59
Biofuels, 22
Biopsies, 32
Black Death, 26
Black holes, 79
Blackouts, 53–54
Bohr, N., 59
Bonding
    atomic theory, 4, 5
    water molecules, 39–40
Brahe, Tycho, 6
Brattain, Walter, 81
Breast cancer, 32
Brownouts, 53–54
Bubonic plagues, 26, 27
Buses, computer data transmission, 86

C

Calcification (of arteries), 30
Calcium channel blockers, 29
Calories in food, 35, 36, 95–96, 103
Cancer, 30–32
    AI and treatment, 88

radiation damage and, 66
radiation therapy, 69
treatment, 32–33
US adult deaths, 27
Capacitive reactance, 51
Capacitors, 51, 84
Carbohydrates, 35, 37, 45, 46, 94
Carbon cycle, 17–18
Carbon dioxide
   carbon cycle, 17–18
   climate change sceptics, 18–20
   greenhouse gas, **14**, 18
   IR absorber, 13
   mitigating global warming, 12, 21–23, 102
   on Venus, 14–15
Carbonic acid, 18
Carbon monoxide, 98
Carbon taxes, 24
Carboxylic acid, 45
Carcinoma, 31
Cardiac arrest, 29–30
Cardiovascular health, 35–36; *see also*
   Heart disease
Carrying capacity, 89, 96
Cellulose, 46
Central processing unit (CPU), 86
Chemical breakthroughs, 38–39
Chemotherapy, 32
Chernobyl, 56
China, 49, 97
Chlorofluorocarbons (CFCs), 11, 38–39
Cholesterol, 46
Chromosomes, 30
Cirrhosis of the liver, 16
Citizen advocacy/power, 103–104
   incentives to engagement, 104–105
Citral, 45
Climate change, 11–25
   carbon taxes, 24
   contributing factor to poor health, 33–34
   electromagnetic (EM) waves and, 12–14
   individual responses to, 25
   political solutions, 24–25
   skeptical views of, 18–21
   solutions to, 21–23
   time to respond, 23–24
   weather events, 23

Collaboration, 10
Colonization, 77, 78
Coma Galaxy, 79
Computed tomography (CT), 30
Computers, 86–87
Condensation model/theory, solar system
   formation, 74–75, 78–79
Conductors, electrical grid, 53, 54
Conservation
   efforts, 21, 103
   engaging public in, 103–104
Control bus, 86
Control rods, 61–62
Control unit, computers, 86
Coronary artery disease, 28–29
Coronavirus, 27, 28; *see also* COVID-19
   pandemic
Coulomb force, 59
Coulombic repulsive forces, 64
Coulombs, 49
COVID-19 pandemic, 24, 27
   response in US 36
   see also mRNA vaccines
Cryotherapy, 33
Current
   in circuits, 49–51
   electronics circuits, 83–85
   transmission lines, 52–54

D

Dark matter, 79–80
Data bus, 86
Data transmission, computers, 86
DC circuit, 49, 50, 84, 85
Dead Sea, 41
Decay, nuclear, 58
DeForest, Lee, 81
Deoxyribonucleic acid (DNA), *see* DNA
Deuterium, 64
Diet, balanced, 35
Diodes, 83–84
Dirty bomb, 69
Disease, 28–32
   climate change and, 16
DNA, 30–32, 46
   Sequencing, 46

Doping, 82–83
Drinking water, 42
Droughts, 15

E

Edison, Thomas, 48
Efficiency, 94–95
  of devices, 105
Einstein, A., 8
Electrical grid, 48–55
  DC and AC circuitry, 49–51
  improvements to, 54–55
  overview, 52
  smart grid, 54
  transmission lines, 53
  vulnerability of, 53–54
Electromagnetic (EM) waves, 12–14
Electronics, 81–89
  AI and decision-making, 88–89
  amplification, 85
  atomic structure, 81–83
  computers, 86–87
  doping, 82–83
  integrated circuits (ICS), 81, 85
  nuclear attacks and devices, 87–88
  optical circuits, 85, 87
  pn junctions, 83–84
  semiconductors, 82
  transistors, 84–85
Electrons
  absorption, 13
  atomic structure, 81–83
  atomic theory, 5
  early atomic models, 56–57
  li-on batteries, 38
  optical circuits and, 87
  pn junctions, 83–85
  QDs and, 39
  solar cells, 97
  spin, 73
  units of charge, 49
  water molecule, 39–40
Endoscope, 32
Energy
  chemical, 92
  compared to work, 91–92
  conservation of, 93–95, 103–104
  electrical, 92

future of, 102–103
geothermal, 102
and human body, 95–96
kinetic, 91
mechanical, 91
nuclear, 92, 102
potential, 91–92
power, 92–93
radiant, 92
sources, 96–98
unit measurement of, 91
Esters, 45
Ethanol, 44
Ethers, 44
Exercise, 35–36
Exoplanets, 76–77
Exploration of universe, extent of, 77–79
External-beam radiation therapy, 33

F

Fermentation, 44
Fermi, E., 60
Firewood, 99
Fission, nuclear, 60–61
Floods, 15
Fluorescent bulbs, 105
Fossil fuels, 18, 22, 43, 48, 90, 96, 98, 102
Freezing, 40
Frequency, 12
  AC circuit, 51
Frost heave, 40
Functional groups, 42, 44, 45
Fusion, nuclear, 63–64, 102

G

Galaxies, 71
  development of, 76
  rotation, 79–80
Galileo, 7
Gasoline, 43–44
Geiger, H., 57
Genes, 30–31
Geoengineering solutions to climate
    change, 22–23
Geothermal energy, 48, 96, 102
Glacial ice, climate change and, 17
Global warming, *see* Climate change
Glucose, 45, 46

Glycerides, 46
Goodenough, John B., 38
Gravitational wave detection, 3–4
Gravity, 6, 7–8
    effects on light, 8
    nuclear fusion and, 63
Great Influenza Epidemic, 26–27
Greenhouse gases, 13–15, **14**, 16, 19,
    21–22, 102
    carbon dioxide and, 18, 20, 98

H

Hahn, Otto, 60
Health, maintenance of good, 35–36
Heart disease, 27–30
Heat, effect on transmission lines, 53, 54
Heat cramps/exhaustion/stroke, 33
Heat pumps 99–100
    coefficient of performance (COP), 100
High blood pressure, *see* Hypertension
HIV/AIDS, 27
Hong Kong flu pandemic, 27
Hormone therapy, 32
Hubble Space Telescope, 79
Hurricanes, 23
Hydrocarbons, 37, 42–45, 96–97
Hydrogen
    amines, 45
    first elements, 71–72
    interstellar space, 72–73
    nuclear fusion, 63–64
    pH scale, 99
    planet formation, 73–74
    water molecule, 39–40
Hypertension, 29
Hyperthermia (cancer treatment), 32
Hypotheses, 1–2

I

Ice, formation of, 40
ICs (integrated circuits), 81, 85
Imaging tests, 31–32
Immunotherapy, 32
Impedance, 50, 51
Implosion bomb, 67
Incentives
    citizen engagement, 104–105
    renewable energy resources, 24–25

Independent-particle (shell) model, 59–60
Indirect gain systems, 101
Inductive reactance, 51
Inertia, 7
Infrared (IR) waves, 12, 13
Internal radiation therapy, 33
International Atomic Energy Agency
    (IAEA), 67
Interstellar space, 72–73
Isomerization, 43

J

James Webb Space Telescope, 77
Joule (J), 91

K

Kepler, Johannes, 6–7
Ketones, 45
Kilowatt (kW)/kilowatt-hours (kWh), 21,
    93, 104
Knowledge, 1–4
    scientific, 8–9

L

Lactic acid, 45
LED lighting, 93, 105
Leukemia, 31, 32
Lipids, 46
Liposarcoma, 31
Liquid drop model, 59, 60
Lithium-ion (Li-ion) batteries, 38
Lymphoma, 31, 32

M

Mars, 15, 74
    colonization of, 78
Marsden, E., 57
Mass-energy conversions, 59
Mental health, 34
Metastasis, 31
Methane, 13, **14**, 16, 43, 96
Methanol, 44
Methodologies, 2–4
Milky Way galaxy, 71
Miller, Stanley, 76
Mining minerals, 77–78
Missing mass problem, 79

Molecular theory of gases, 5
Molecules
  absorption, 13
  carbohydrate compounds, 46
  development of more complex, 76
  global warming and expansion of, 17
  molecular theory of gases, 4–5
  nucleic acids, 46
  organic, 42–44
  radiation damage, 64–65
  water, 39–40
Montreal Protocol, 39
Moon, colonization of, 78
Morphine, 45
mRNA vaccines, 28
Muscles, 91
Myeloma, 31

N

Natural gas, 43, 48, 96–97
Natural selection, 76
Newton, I., 7
  second law, 4
  third law, 5
Nicotine, 45
9/11 attacks, 68
North Carolina, electrical grid
    vulnerability, 53
Npn bipolar junction transistor, 84
Nuclear energy
  background, 56–57
  dirty bomb, 69
  fission, 60–61
  fission reactors, 61–63
  fusion, 63
  fusion reactors, 63–64
  future of, 102
  leakage of radioactive materials, 67–68
  models of nuclei, 59–60
  nuclear power plants, 69–70
  physics of, 57–58
  radiation therapy, 69
  radioactivity and humans, 64–66
  storage of waste in US, 68, 70
  weapons, 67
Nuclear power plants, 22, 69–70
Nuclear waste, 62
  storage offsite, 68

Nuclei 58
  models of, 59–60
Nucleic acids, 46
Nucleons, 58
  liquid drop model, 59
Nucleotides, 46

O

Ohm, 50
Optical circuits, 85
  primary advantages, 87
Organic molecules, 42–46
Oxygen, body consumption of,
    36
Ozone, 11, 34, 38–39

P

Pandemics, 26–27; see also COVID-19
    pandemic
Particulates, 99
Passive solar houses, 100–101
Pests, climate change and, 16
Petroleum, 37, 43–44, 96, 102
pH scale, 99
Phenols, 44
Photodynamic therapy, 32–33
Photosynthesis, 37, 92, 94
Plague of Justinian, 26
Planetary model of atoms, 57
Planetary orbits, 6–7, 78
Planetesimals, 74
Plant-based diets, 103
Plasma, 64
"Plum pudding model", 56–57
Plutonium bomb, 67
pn junctions, 83–84
Poison ivy, 44
Polar vortexes, 39
Potential difference, 49
Power, 92–93
  worldwide consumption,
    22
Power plants, 93–95
Precipitation, 41–42, 98
Proteins, 45–46
Protons, 58
Protoplanets, 74

Q

Quantum dots (QDs), 39
Quarks, 58

R

Radiation damage
    to electronics, 87–88
    to humans, 64–66
Radiation therapy, 33, 69
Radioactive drugs, 33
Radioactive waste, 62
RBE (relative biological effectiveness)
    factor, 65
Reactor core, 61–62
Reactor safety, 62–63, 67–68
Relativity, 8
Renewable energy, 48–49, 102–103
Reproduction constant K 61
Resistance of circuit, 49–50
Ribonucleic acids (RNAs), 46
Rotation, galaxy rates of, 79
Rubin, Vera, 79
Russian Flu pandemic, 27
Rutherford, E., 57

S

Sarcoma, 31
Schwarzschild radius, 79
Scientific questions, 1–2
Scientific theories, 2
    atomic theory of matter, 4–5
Sea ice, climate change and, 17
Semiconductors, 82
SETI (Search for Extraterrestrial
    Intelligence), 77
Sewage treatment, 42
Shell model, see Independent-particle
    (shell) model
Shockley, William, 81
Skepticism, 9
    views on climate change, 18–21
Skin cancer, 31, 33, 38
Smart grid, 54
Software, 86
Solar activity, global warming and, 19
Solar energy/power, 22, 97
Solar irradiance, 19

Solar system, models of origin, 73–75, 78
Space X, 77
Spin states, 60, 73
Stents, 29
Steroids, 32, 46
Strassman, Fritz, 60
Strong nuclear force, 58
Sugars, 46
Sulfur, 98
Sun
    modifying climate, 22–23
    nuclear fusion, 63, 64
    planetary orbits around, 6–7, 78
    solar cycles, 19
    solar heating, 100–101
    source of energy, 90, 95, 97
    Venus position to, 14, 15
Swift, Taylor, 24
Swine Flu pandemic, 27
System software, 86
Systemic radiotherapy, 33

T

Target heart rate, 35
Texas, grid infrastructure, 55
Tobacco use, 36
Transformers, 52
Transistors, 84–85
Transmission lines, electrical grid, 53
    effect of heat on, 53, 54
    improvements to, 54–55
Transpiration, 41
Triglycerides, 46
Trust, citizens in science/scientists, 9
Tyndall, John, 14

U

Ultraviolet (UV) radiation, 38–39
    fluorescent bulbs, 105
Universe
    biochemical molecules, 76
    black holes, 79
    colonization, 77, 78
    dark matter, 79–80
    expanding, 71
    exploration of, 77–79
    harvesting valuable materials, 77–78

interstellar space, 72–73
life forms, 76–77
origins of solar system, 73–74, 78–79
star formation, 75–76
Uranium, 57–58
  bomb, 67
  enriched, 61, 67
  fission, 60–61
  world-wide reserves, 69–70
Uranus, 74, 75
Urey, Harold, 76
USB (Universal Serial Bus), 86

V

Vaccines, 27
  mRNA, 28
Valence electrons, 82–83
Vanilla, 45
Ventricular fibrillation, 29–30
Venus, greenhouse effect on, 14–15

W

Water
  cycle, 33–34, 41–42
  extreme weather and supply of, 33–34

heaters, 100–102
high specific heat, 41
hydrosphere, 39–42
ice formation, 40
precipitation, 41–42, 98
purification, 42
sewage treatment, 42
as source of energy, 98
surface tension, 41
Watt, James, 92
Wave speed, 12
Wavelength, 12
  light scattering, 72
Weapons, nuclear, 67
Weather events, climate change and,
  16, 23
Whittingham, M. Stanley, 38
Wildfires, 15–16
Wind power, 22, 97

Y

Yoshino, Akira, 38

Z

Zwicky, Fritz, 79